TEUBNER-TEXTE zur Physik Band 35

R. Luzzi / A. R. Vasconcellos / J. G. Ramos

Statistical Foundations of Irreversible Thermodynamics

TEUBNER-TEXTE zur Physik

This regular series includes the presentation of recent research developments of strong interest as well as comprehensive treatments of important selected topics of physics. One of the aims is to make new results of research available to graduate students and younger scientists, and moreover to all people who like to widen their scope and inform themselves about new developments and trends.

A larger part of physics and applications of physics and also its application in neighbouring sciences such as chemistry, biology and technology is covered. Examples for typical topics are: Statistical physics, physics of condensed matter, interaction of light with matter, mesoscopic physics, physics of surfaces and interfaces, laser physics, nonlinear processes and selforganization, ultrafast dynamics, chemical and biological physics, quantum measuring devices with ultimately high resultion and sensitivity, and finally applications of physics in interdisciplinary fields.

Statistical Foundations of Irreversible Thermodynamics

By Prof. Dr. Robert Luzzi
Prof. Dr. Áurea Rosas Vasconcellos
Prof. Dr. José Galvão de Pisapia Ramos
State University of Campinas
UNICAMP, São Paulo

B. G. Teubner Stuttgart · Leipzig · Wiesbaden

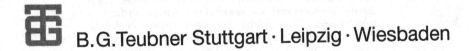

Prof. Dr. Roberto Luzzi

Born in 1936 in Buenos Aires. M. Sc. in 1962, Ph. D. in Physics in 1966. Post-Doctoral Associate at Purdue University (USA). Assistant Professor at the University of Southern California (USA). 1976 John Simon Guggenheim Memorial Foundation Fellow. Since 1971 Professor of Physics at the State University of Campinas (São Paulo, Brasil).

Prof. Dr. Áurea Rosas Vasconcellos

Born in 1939 in Recife. Ph. D. in Physics in 1976. Visiting Scientist at the University of California (USA, 1977–1978). Presently Associate Professor of Physics at the State University of Campinas (São Paulo, Brasil).

Prof. Dr. José Galvão de Pisapia Ramos

Born in 1938 in São Paulo. Ph. D. in Physics in 1970. Assistant Professor at the University of São Paulo (1966–1971). Visiting Scientist at the Massachusetts Institute of Technology (USA, 1974–1976). Associate Professor of Physics (1971–1991) and after retirement Invited Professor at the State University of Campinas (São Paulo, Brasil).

Die Deutsche Bibliothek – CIP-Einheitsaufnahme
Ein Titeldatensatz für diese Publikation ist bei
Der Deutschen Bibliothek erhältlich.

1. Auflage September 2000

Der Verlag Teubner ist ein Unternehmen der Fachverlagsgruppe BertelsmannSpringer.

www.teubner.de

Umschlaggestaltung: Peter Pfitz, Stuttgart

ISBN 978-3-519-00283-3
DOI 10.1007/ 978-3-322-80019-0

ISBN 978-3-322-80019-0 (eBook)

To my parents in memoriam
To Edgardo, and Roman and Marielba (RL)

To my children: Vanessa, André, and Aline,
and to Denise in memoriam (ARV)

To my mother and her parents in memoriam
To Watfa, Daniela and Patricia (JGR)

Prologue

It is considered that at the present time — the end of the twentieth century and of the second milenium — there are four pillars which give sustainment to the theoretical aspect of Physics. They are General Relativity, Quantum Mechanics, Elementary Particle Theory, and Statistical Physics. We are here interested in the latter.

Paraphrasing Lawrence Sklar [1], Thermodynamics and Statistical Mechanics are a pair of fascinating theories. Originally restricted to give account of the macroscopic thermal behavior of matter, these branches of physics are nowadays in an equal foot with the other three fundamental components, cited above, of our present scientific point of view of the world. More precisely, they are an indispensable supplement for the other three: Suffice to recall that the Biosphere, and in it we as a macroscopic system together with all the other living creatures with their origin, evolution, and functioning, can only be analized in terms of thermodynamics — particularly in what constitutes the case of open dissipative systems out of equilibrium — together with the connection with microscopic mechanics (classical or quantal) provided by statistical mechanics.

Altogether, thermal phenomena, magnetic phenomena, electro-magnetic phenomena in material media, the structure of matter in all its phases, the spectral distribution, all experiment in the laboratory, and so on and so forth, require a description and an explanation in the realm of thermodynamics and statistical mechanics.

Summarizing, even though Thermodynamics was originally formulated to provide the general laws which would govern the thermal properties of matter and the interplay between heat and mechanical work, it has evolved to become a more general theory having an extraordinarily ample scope. On the other hand, we do have Statistical Mechanics, whose founding fathers had originally in mind a tentative to obtain an understanding of the thermal phenomena, with the intention to avoid the necessity to invoke laws and principles purely thermodynamic independent of those of the microscopic world of the mechanics of atoms and molecules. However, such hope to eliminate those purely thermodynamical principles in the description of Nature have been constantly thwarted. It appears to be an inevitable necessity, at the more profound levels in the development of ideas,.to invoke some hypotheses which cannot be derived from the basic laws of kinematics or of the dynamics of many-particle systems.

At this point it was unavoidable to introduce a new and remarkable idea in Physics, namely laws of a probabilistic character. On this we cite Jacob Bronowski [2], who stated that "This [the ideia of chance] is the revolutionary thought in modern science. It replaces the concept of the *inevitable effect* by that of the *probable trend.* Its technique is to separate so far as possible the steady trend from local fluctuations [...] The future does not already exist; it can only be predicted."

Moreover, Richard Feynman [3] maintained that "nature appears to have probability in it. It seems to be somehow intrinsic. Someone has said it this way — 'Nature herself does not even know which way [a physical system] is going to go'." James C. Maxwell also commented that "The true logic for this world is the calculus of probability, which takes into account the value of the probability which is present, or should be, in the man's rational mind".

It is worth noticing that, differently to the disciplines of general relativity or quantum mechanics, statistical mechanics is not facing us as a unique theory, firmly formulated and established, but — at least up to the present moment — is constituted by an aggregate of approaches and formalism which are associated to different schools of thought. Although some aspects of the theory are reasonably well understood and have ample acceptance, some crucial aspects like a proper description of the nonequilibrium thermodynamic state of many-body systems, the introduction of irreversibility (the famous "time's arrow", so called by Sir Arthur Eddington), a proper definition of entropy and its relation to the underlying microscopic mechanics, etc., are subjects of intense, controverted, sometimes rispid and unending discussions.

In this book we deal with some aspects of the physics of many-body systems driven away from equilibrium, that is, the characterization and irreversible evolution of their macroscopic state. In the following Chapters, first we briefly, and partially, consider aspects of the present status of phenomenological irreversible thermodynamics and nonequilibrium statistical mechanics. Recently several aspects, concepts, and ideas in the statistical physics and thermodynamics of nonlinear nonequilibrium systems have been discussed in a relevant set of articles published in the namesake book [4], to which we call the attention of the reader: the present book may be considered as a complement to that book by partially touching upon the question of the statistical foundations of irreversible thermodynamics on the basis of a Gibbs-like ensemble approach for nonequilibrium (and then dissipative) systems. After short comments on Classical Irreversible Thermodynamics, its conceptual and practical shortcomings are pointed out, as well as the efforts undertaken to go beyond its limits, consisting of particular approaches to a more general theory of Irreversible Thermodynamics. In particular, a search for statistical-mechanical foundations of Irreversible Thermodynamics, consisting in building a statistical thermodynamics, is based on the Nonequilibrium Statistical Operator Method. This important theory for the treatment of phenomena at the macroscopic level is based on a microscopic molecular description in the context of a nonequilibrium ensemble formalism. We draw attention to the fact that this method may be considered to be encompassed within Jaynes' Predictive Statistical Mechanics and based on the principle of maximization of informational entropy. Finally, we describe how, in fact, the statistical method provides foundations to phenomenological irreversible thermodynamics, thus giving rise to what can be referred to as Informational Statistical Thermodynamics. It is included an Appendix with a chronology of relevant events in the history of Thermodynamics and Statistical Mechanics.

We conclude this Prologue with several due acknowledgments. We thank the different Agencies and Foundations which, at one time or another, provided finantial support to our Research Group at the Institute of Physics of Unicamp, namely, the São Paulo State Research Foundation (FAPESP), the Brazilian National Research Council (CNPq), the Ministry of Planning (Finep), Unicamp Foundation (FAEP), IBM-Brasil, and the John Simon Guggenheim Memorial Foundation (New York, USA). Useful and enlightening discussions with Leopoldo García-Colín (UAM — Mexico) and José Casas-Vázquez and David Jou (UAB — Catalunya,

Spain) are gratefully avowed. We have also benefited from the work of graduate students and post-doctorals, and particularly we thank Marcus Vinicius Mesquita (presently at Humboldt Universität, Berlin), Sergio A. Hassan (presently at the Mount Sinai School of Medicine, CUNY, New York) and Justino R. Madureira (presently at Rostock Universität, Rostock). We thank also Marcus Vinicius Mesquita for the typesetting of the manuscript. We also thank the clerical staff at our Institution, mainly the Secretaries of the Department and Institute, and the Head Librarian and her assistants, for the administrative support provide to our Group.

Campinas, November 1999

R. Luzzi

A. R. Vasconcellos

J. G. Ramos

Contents

Introduction

Thermodynamics is considered to be an offshoot of the Industrial Revolution that began in England in the second half of the 18th Century and from there spread to other parts of the world. The word thermodynamics is derived from the Greek *thermé* (meaning heat) and *dynamis* (meaning force). As well known, the origins of thermodynamics are founded in the early 19th century in the study of the motive power of heat; that is, the capability of hot bodies to produce mechanical work. However, there are of course precursors to these ideas: Temperature is probably the earliest thermodynamic concept to attain operational status (early in the 17th century with Galileo). The science of calorimetry beginning in the late 18th century (contemporary with the beginning of the Industrial Revolution) led to the establishment of the caloric theory of heat [5]. Clausius in the second half of the 19th century established Thermodynamics as a clearly defined science. The connection of Thermodynamics with Mechanics is first achieved through kinetic theory with the work of D. Bernouilli, J. Herapath, J. J. Waterston, R. Clausius, J. C. Maxwell, and finally L. Boltzmann, later through *Statistical Mechanics*, whose main purpose is to determine the thermodynamic properties and values of macroscopic observables in terms of the dynamical laws that govern the motion of the constitutive particles of the system. It is not easy to establish precisely the dates of the birth of Statistical Mechanics. Its founding fathers (scientists of the 19th century) are considered to be J. C. Maxwell, L. Boltzmann and J. W. Gibbs. Gibbs provided a mathematical method so orderly and systematic that it constitutes the foundation of this science. It is fundamentally based on probability theory, mainly in the spirit introduced by Laplace [6]. As stated by J. Bronowski [2]: "This is the revolutionary thought in modern science. It replaces the concept of *inevitable effect* by that of *probable trend.* Its technique is to separate so far as possible the steady trend from local fluctuations [...]. The future does not already exist; it can only be predicted". Preliminaries to statistical mechanics can be considered to go back to the fifth century before Christ with the first atomists — at least the first so recorded — among the philosophers of the ancient Greek city states, mainly Democritos. In the Appendix I we present a tentative chronology of the fundamental events in the history of Statistical Mechanics and Thermodynamics, commencing sometime around the year 400 before Christ to near the present day, accompanied by a bibliographical list of some of the seminal papers.

Thermodynamics as well as Statistical Mechanics can be separated into two parts, namely, that corresponding to the study of systems in equilibrium, and that corresponding to nonequilibrium situations. The latter case will be referred to as Irreversible Thermodynamics, and two distinct important cases can be characterized: The so-called *linear regime* and the *nonlinear regime.* Nonlinear Irreversible Thermodynamics is, evidently, a branch of the nowadays fashionable *Nonlinear Physics.* More precisely it can be considered as a particular case of the emerging *Theory of Complex Dynamical Systems* [7–9] in those aspects related to self-organization in open dissipative systems far from equilibrium, and its possible relevance in the emergence, functioning and evolution of life. Ilya Prigogine (the 1977 Nobel Prize) has extensively dealt with this new paradigm, complexity, being one of its founding fathers. Comments upon this question of dissipation and order and its possible statistical-mechanical foundations had been made in reference [10].

In the present book we consider a general overview of Nonlinear Irreversible Thermodynamics and its statistical foundations, based on the maximization of the informational-statistical entropy and Bayesian methods [11–13], leading to what can be termed *Informational Statistical Thermodynamics* [14]. This has been done on the basis of an extension of a former review article, namely "On the Statistical Foundations of Irreversible Thermodynamics", by the same authors of this book, which has been published by Fortschritte der Physik/Progress of Physics, a publication of Akademie Verlag (Berlin).

Chapter 1

Irreversible Thermodynamics

Let us recall that nonequilibrium thermodynamics is a field theory at a macroscopic and phenomenological level, which deals with states and processes in systems lying beyond equilibrium (either by a large amount, or close to equilibrium), with equilibrium being characterized as the state where there are no temporal modifications of the state variables and the fluxes of dynamical quantities through the frontiers of the system vanish. Nonequilibrium thermodynamics is mainly related to transport and relaxation phenomena and steady states in continuum media, as well as to how the steady states are obtained, their stability, and the accompanying relaxation phenomena.

The case of equilibrium thermodynamics (sometimes referred to as Thermostatics) deals basically with reversible processes, namely infinitesimal and quasi-static processes. On the other hand nonequilibrium thermodynamics is involved with finite and irreversible processes, and, for that reason, it is also denominated, as noticed, Irreversible Thermodynamics. As a general rule special interest is currently focused on the case of open systems, which are coupled to external sources which supply them energy and matter. This manifests itself in macroscopic changes of the thermodynamic variables which naturally become dependent on space position and on time.

The principal characteristic of the processes involved, whether they are stationary or time dependent, is that they evolve with a positive production of internal entropy. Here we make contact with the second fundamental principle of thermodynamics. According to Planck [15], "The second law of thermodynamics is essentially different from the first law, since it deals with a question in no way touched upon by the first law, viz., the direction in which a process takes place in nature". The second law has several equivalent formulations: the one due to Clausius refers to heat conduction, and, again according to Planck [15], it can be expressed as "heat cannot by itself pass from a cold to a hot body. As Clausius repeatedly and expressly pointed out, this principle does not merely say that heat does not flow from a cold to a hot body — that is self-evident, and is a condition of the definition of temperature — but it expressly states that heat can in no way and by no process be transported from a colder to a warmer body without leaving further changes, i.e. without compensation".

Hence, as we have said, nonequilibrium thermodynamics is intrinsically linked to the concept of irreversibility, which is completely beyond the scope of the laws of mechanics, let the latter be classical or quantal, relativistic or not. The concept of irreversibility appears clearly in Clausius' work, who, as already noted, can be considered the founder of Thermodynamics as an autonomous and unified science. His presentation of the second law appears precisely as a criterion for evolution, governing irreversible behavior. It implies the definition and use of the extremely difficult concept of entropy. The word is said to derive from the Greek *εντρωπη*, meaning evolution. Tradition has it that Clausius considered *ητροπη*, meaning change and added the prefix *en* to make it similar to energy. It is also interesting to notice that there exists the word *εντρεπομαι*, meaning tendency to change in a given direction. Using the usual notation, for an infinitesimal process the state function entropy S suffers a modification $dS = \delta Q/T$ if the process is reversible, while for irreversible processes there follows $dS > \delta Q/T$, where δQ is the heat exchange in the process and T the absolute temperature, and δ stands for a nonexact differential (thus, $1/T$ acts as an integrating factor). The last inequality can be expressed in the form of the balance equation

$$dS = \frac{\delta Q}{T} + \frac{\delta Q'}{T}, \qquad (1.1)$$

with $\delta Q' > 0$. In this form of the second law, the quantity $\delta Q'$, called by Clausius *uncompensated heat*, can be considered as a first evaluation of the degree of irreversibility of a natural process. It encompasses the case of isolated systems in equilibrium when $\delta Q = 0$, and $dS = 0$ (*i.e.* $\delta Q' = 0$), that is, when is reached the state of maximum entropy. In

Eq. (1.1) S is the equilibrium entropy obtainable through calorimetric measurements.

Equilibrium thermodynamics describes states of matter that are greatly privileged: Planck [15] has emphasized that the second law distinguishes among the several types of states in nature, some of which act as atractors to others: Irreversibility is an expression of this attraction. For systems together with the reservoirs to which they are connected there is an attraction towards equilibrium (thermodynamic potentials at a minimum value compatible with the constraints imposed by the reservoirs). In nonequilibrium systems, while in a stationary state in a linear regime near equilibrium, the attractor is the state of minimum internal entropy production (as shown on Section 4.9, p. 90).

Nonequilibrium systems evolve in time accompanied by irreversible dissipation, and, in open systems there is, in general, a flux of entropy through their boundaries. It has been stated [16] that nobody is able to live without the sensation of the flux of entropy, that is, of that sensation that under diverse forms regulate the activities related to the maintenance of the organism. In the case of mammals this sensation includes not only the sensations of cold and warm, but, also, the pangs of hunger or the satisfaction after a meal, the sensations of being tired or rested, and many others of the same type.

According to Eq. (1.1), the change of entropy is composed of two separate types of contributions: the term $\delta Q/T$ related to the exchange of heat with the surroundings, expressed as d_eS, and the term $\delta Q'/T$, due exclusively to the irreversible processes that develop in the interior of the system, expressed as d_iS so we can write

$$dS = d_eS + d_iS, \qquad \text{with} \qquad d_iS \geq 0\,. \tag{1.2}$$

Equation (1.2) provides the framework for an entropy equation of balance. Let us emphasize that the term d_eS involves all the contributions resulting from exchanges (energy, matter, etc.) with the environment.

It is certainly a truism to say that entropy has a very special status in physics, expressing, in a very general way, the tendency of physical systems to evolve in an irreversible way, characterizing the eventual attainment of equilibrium, and given a kind of measurement of the order that prevails in the system. It is important to stress that it is a very well established concept in equilibrium situations, however requiring an extension and clear comprehension in the case of open systems, mainly in far-from-equilibrium conditions, a question not satisfactorily solved as yet. On the basis of the use of entropy as a state function, the properties of systems in *equilibrium* are very well described when in conjunction with the two fundamental laws.

In *nonequilibrium* situations, in contrast to equilibrium thermodynamics which constitutes an undisputed universal discipline, the thermodynamics of irreversible processes has not achieved a phenomenological formulation and a methodology that can be considered satisfactory. Let us recall that nonequilibrium thermodynamics has quite wider scope aims, and also a greater complication, than those of equilibrium thermodynamics, beginning with the fact that it needs to describe the changes in space and time in systems subjected to mechanical and thermal perturbations.

It is a known fact that the fields of nonlinear nonequilibrium thermodynamics and the accompanying kinetic and statistical theories have been the object of large interest in recent decades. The field has as yet not attained a satisfactory definitive level of description and, then, it is natural that it be the subject of intense and lively discussion and controversy (for example, on one particular aspect, see Letters Section in Physics Today, November 1994, pp. 11–15 and 115–117). A first, and a quite successful (within its domain of validity) approach to irreversible thermodynamics was Classical (sometimes referred to as Linear or Onsagerian) Irreversible Thermodynamics (CIT; as described, for example, in the already classic books of references [17] and [18]), built largely on the basis of the fundamental work of Lars Onsager [19]. During the fifties and sixties it was carried beyond the linear regime — in the sometimes referred-to as Generalized Classical Irreversible Thermodynamics — by Ilya Prigogine and the so-called Brussel's School [20].

To encompass arbitrarily far-from-equilibrium situations phenomenological Classical Irreversible Thermodynamics is being superseded by new attempts. It is considered that there exist four approaches to the description of Thermodynamics [21], namely:

(i) The one based on the two fundamental laws of thermodynamics and the rules of operation of the Carnot cycles. This is sometimes referred to as the engineering point of view or the C-K Thermodynamics (for Clausius and Kelvin);

(ii) The mathematical approach, as the one based on differential geometry and topology instead of Carnot cycles, sometimes referred to as the C-B Thermodynamics (for Caratheodory and Born);

(iii) The axiomatic point of view, replacing Carnot cycles and differential geometry by a set of basic axioms, that try to encompass the previous ones and extend them, which may be referred to as T-C Thermodynamics (for Tisza and Callen);

(iv) The statistical-mechanical point of view, based of course on the substrate provided by the microscopic mechanics (at the molecular, or atomic, or particle level) plus theory of probability, which can be referred to as Gibbsian Thermodynamics or Statistical Thermodynamics.

It is not an easy task to readily classify all existing attempts within this scheme. Among several approaches we can mention — and we apologize for those omitted —, *Rational Thermodynamics*, as proposed by Truesdell [22]; a *Geometrical-Topological approach*, as an extension and generalization of Caratheodory's method, by P. T. Landsberg [23]; what we call *Orthodox Irreversible Thermodynamics*, as proposed by B. Chan-Eu and followers [24]; *Extended Irreversible Thermodynamics*, originated and developed by several authors and largely systematized and improved by J. Casas-Vázquez, D. Jou, and G. Lebon of the so-called Catalan School of Thermodynamics [25]; a *Generalized Kinetic approach* developed by L. S. García-Colín and the so-called Mexican School of Thermodynamics [26]; the *Wave approach to Thermodynamics* due to I. Gyarmati [27]; the approach so-called *Generics* by M. Grmela and followers [28]; the *Holotropic approach* by N. Bernardes [29]; *Informational Statistical Thermodynamics* (or Information-theoretic Thermodynamics) with mechanical statistical foundations, initiated by A. Hobson [30] and whose systematization and extension are described here.

We may say that Rational Thermodynamics, the Topological approach and Generics belong to level (ii); Orthodox Irreversible Thermodynamics to level (i); Extended Irreversible Thermodynamics to level (iii); Holotropic Thermodynamics also to level (iii); Informational Statistical Thermodynamics, evidently, to level (iv).

Let us briefly consider next, the cases of Classical Irreversible Thermodynamics (CIT), Rational Thermodynamics (RT), and Extended Irreversible Thermodynamics (EIT).

In the case of *Classical Irreversible Thermodynamics*[17, 18] we can summarize its basic tenets as:

CIT.1) The local equilibrium hypothesis: in each material point of the system (macroscopic in the sense of taking an infinitesimal element of volume — an elemental cell — around space point $\boldsymbol{r}$ however containing a large number of particles), the thermodynamic state is described by the local densities of the variables that are used in equilibrium. Typically they are the energy $u(\boldsymbol{r},t)$, the specific volume $\varrho^{-1}(\boldsymbol{r},t)$ (where $\varrho(\boldsymbol{r},t)$ is the density of matter), and the concentration of the different chemical components, $n_j(\boldsymbol{r},t)$, $j = 1, 2, \ldots, s$. It should be noticed that the concept of local equilibrium implies a dependence on time in the basic variables that is not instantaneous, but characterizing a time interval Δt sufficiently long as to ensure, in the neighbourhood of point $\boldsymbol{r}$, that enough interparticle collisions have occurred to establish the postulated local equilibrium. More precisely, the hypothesis requires that in the elemental cell around $\boldsymbol{r}$ the change on the extension of a mean free path l of the thermodynamic variables u, ϱ^{-1}, n_j [let us generically call them $Q_j(\boldsymbol{r},t)$] be small, that is, it must follow that

$$\frac{l\left|\nabla Q_j(\boldsymbol{r},t)\right|}{\left|Q_j(\boldsymbol{r},t)\right|} \ll 1\,, \tag{1.3}$$

meaning near uniformity within the elemental cell. It requires also that temporal changes in the mean interval between collisions τ be also small, i.e.

$$\frac{\tau\left|\partial Q_j(\boldsymbol{r},t)/\partial t\right|}{\left|Q_j(\boldsymbol{r},t)\right|} \ll 1\,. \tag{1.4}$$

Finally, the macrovariables must be normal, that is, their fluctuations $\Delta^2 Q$ must be negligible when compared with their local values, i.e.

$$\frac{\left|\Delta^2 Q_j(\boldsymbol{r},t)\right|}{\left|Q_j(\boldsymbol{r},t)\right|^2} \ll 1\,. \tag{1.5}$$

CIT.2) A local nonequilibrium entropy η is defined, such that — similarly to the equilibrium case — it satisfies the Pfaffian-like form (a Gibbs relation in each elemental cell)

$$T(\boldsymbol{r},t)\,d\eta(\boldsymbol{r},t) = du(\boldsymbol{r},t) + p(\boldsymbol{r},t)\,d\varrho^{-1}(\boldsymbol{r},t) - \sum_{j=1}^{s} \mu_j(\boldsymbol{r},t)dn_j(\boldsymbol{r},t)\,, \tag{1.6}$$

which introduces local intensive variables at time t, consisting of $T(\boldsymbol{r},t)$, $p(\boldsymbol{r},t)$, and $\mu_j(\boldsymbol{r},t)$. They are related to the differential coefficients of the entropy, representing local temperature, pressure, and chemical potentials, similarly to the equilibrium case. Furthermore, the local entropy satisfies a balance equation of the form

$$\frac{\partial}{\partial t}\eta(\boldsymbol{r},t) + \text{div}\,\boldsymbol{I}_\eta(\boldsymbol{r},t) = \sigma_\eta(\boldsymbol{r},t)\,, \tag{1.7}$$

where $\boldsymbol{I}_\eta$ is the entropy flux, and σ_η the local production of entropy due to sources and sinks, with both quantities to be determined in each case.

CIT.3) Expression (1.7) is a kind of local version of Clausius uncompensated heat, but, following Eq. (1.2), admitedly valid locally and at each time t. Hence, the production of entropy due to the dissipative effects that develops in the system satisfies that

$$\sigma_\eta(\boldsymbol{r},t) \geq 0\,. \tag{1.8}$$

CIT.4) To close the theory it is necessary to incorporate the so-called constitutive equations (phenomenological relations) that connect basic variables and fluxes. Typical examples are Fourier's and Fick's relations respectively given by

$$\boldsymbol{I}_q(\boldsymbol{r},t) = -\kappa\,\nabla T(\boldsymbol{r},t)\,, \tag{1.9a}$$

$$\boldsymbol{I}_n(\boldsymbol{r},t) = -D_n\,\nabla n(\boldsymbol{r},t)\,. \tag{1.9b}$$

Equation (1.9a) establishes a linear relation between the heat flux and the gradient of temperature, with coefficient κ being the thermal conductivity. Moreover, in Eq. (1.9b) we have the relation between the flux of matter (related to the local linear momentum) and the gradient of concentration, where D_n is the diffusion coefficient (It should be noticed that these equations neglect non-local and memory effects). Using Eqs. (1.9), as known, one can derive Fourier equation for the diffusion of heat and

Fick equation for the diffusion of matter, which in the absence of sources are

$$\left[\frac{\partial}{\partial t} - D_q \nabla^2\right] T(\boldsymbol{r}, t) = 0\,, \tag{1.10a}$$

$$\left[\frac{\partial}{\partial t} - D_n \nabla^2\right] n(\boldsymbol{r}, t) = 0\,. \tag{1.10b}$$

with D_q and D_n being the corresponding diffusion coefficients. In general, for a generic set of fluxes $\boldsymbol{I}_k$ and thermodynamic forces $\boldsymbol{F}_k$ [like the gradients in Eqs. (1.9)] it follows that

$$\boldsymbol{I}_k = \sum_{\ell} L_{k\ell} \boldsymbol{F}_\ell\,, \tag{1.11}$$

where the matrix of kinetic coefficients L satisfies the symmetry relations established by Onsager [19], the so called *Onsager's reciprocity relations*. This property is a very relevant result in irreversible thermodynamics, which brought the phenomenological approaches into contact with the microscopic theory of nonequilibrium fluctuations.

Another relevant result in CIT is *Prigogine's theorem of minimum entropy production* [20, 31]. In Thermostatics equilibrium situations are characterized by thermodynamic potentials, whose extremal value in equilibrium acts as an attractor for which the thermodynamic evolution of the system tends irreversibly. According to Prigogine's theorem, in CIT there also exist a function whose variational properties characterize the nonequilibrium stationary state towards which the system evolves: this is determined by the minimum internal production of CIT-entropy compatible with the constraints imposed on the system. This principle implies in a kind of "inertia": when the external constrains do not allow the system to attain equilibrium it does "the best it can" by evolving towards the state of minimum entropy production, so we may say that it goes to the state as near to equilibrium "as possible". Hence, CIT predicts the behavior of macroscopic systems pointing to the fact that they tend to a steady state with a minimum of activity compatible with the fluxes on which it feeds. This implies that along such evolution the initial conditions are forgotten: whichever they are the system tends to a determined steady state under the imposed constraints, and, as a consequence, the reaction of the system to any change through the boundary conditions is completely predictable.

What are the difficulties and limitations inherent to CIT? In first place it is clear that it applies only to situations near equilibrium (linearity in Onsager's relations). Also, in certain situations the theory is not able to provide a satisfactory agreement with experimental results, and then verification fails. Furthermore there exist conceptual difficulties associated to the fact that the constitutive equations [v.g. Eqs. (1.9)] lead to diffusion-like equations [Cf. Eqs. (1.10)] that are partial differential equations of the parabolic type and, therefore, they are associated to propagation of thermal perturbations with infinite velocity, a clearly uncomfortable characteristic.

Failures and limitations of CIT, particularly the nowadays growing interest in the inclusion of studies of evolution and steady states in systems arbitrarily away from equilibrium (i.e. outside Onsager's linear regime) evidently requires to look for formalisms that would permit to enlarge the domain of validity of CIT.

We have already mentioned two approaches, RT and EIT. In Rational Thermodynamics [22] the concept of local equilibrium is abandoned and the concept of memory is introduced: the behavior of the system at time t is determined by the values of the macrovariables taken during the previous history of evolution. The thermodynamic state of the system is described by response functionals, which, however, present certain difficulties in being related to observables (critical analyses are due to, among others, Meixner [32], Müller [33], Rivlin [34]; we do not go here into details on RT; for a complete description see ref. [22])

Another approach is *Extended Irreversible Thermodynamics*. Its objectives are:

I. To generalize CIT going beyond the local equilibrium hypothesis by prying into the shorter wavelengths and smaller time scales phenomena;

II. To obtain the best possible accord with experiment and with the best founded aspects of modern kinetic theory and Statistical Mechanics;

III. To introduce a formalism as simple as possible but built upon a rigorous mathematical framework, and, in particular, to do away with the inconvenient concept in CIT associated with the propagation of thermal perturbations at infinite speed.

EIT has a kind of anticipation in an 1867 article by James Clerk Maxwell [35], who incorporated a term accounting for relaxation in the constitutive equations of ideal gases. It reads

$$-\theta_r \frac{\partial}{\partial t} \overset{o}{\tau} = \overset{o}{\tau} + 2\xi \left(\nabla \overset{o}{:} \boldsymbol{u} \right)^s + \cdots , \tag{1.12}$$

where $\overset{o}{\tau}$ is the traceless part of the symmetric viscous tensor, ξ is the shear viscosity, $\boldsymbol{u}$ is the field velocity, the index s stands for the symmetric part of the tensor, and θ_r is Maxwell's relaxation time. Only recently such an idea was extended by Cattaneo [36] and Vernotte [37]. It acquires a rigorous formulation in EIT, through mainly the work of Nettleton [38–42], Müller [33,43–45], Gyarmati [27], Lambermont and Lebon [46], Jou et al. [47,48], Casas-Vázquez and Jou (of the so-called Catalan School of Thermodynamics responsible for a large systematization of EIT) [49], Lebon [50,51], García-Colín et al., [26,52–55], and others. Reviews of the theory are found in Müller and Ruggieri [45]; Jou et al. [56], García-Colín [57,58], and Lebon et al. [59], while in reference [60] is presented an overview of recent published work; see also reference [61].

A different approach to EIT is adopted by L. S. García-Colín and coworkers, who disagree with the physical interpretations of, basically, entropy and temperature. These authors maintain that their views are supported not only macroscopically but also from kinetic theory and statistical mechanics [26,52,55,57,58].

The basic propositions in EIT are:

EIT.1) Gibbs space (that is the space defined by the thermodynamic variables), also called the space of thermodynamic states, is enlarged in relation to CIT by the inclusion as basic variables of the fluxes and all other quantities deemed appropriate for the better description of the system under consideration.

EIT.2) A nonequilibrium-like entropy is introduced, η, that is, a state function defined on the extended Gibbs space, such that

$$\Theta(\boldsymbol{r},t)d\eta(\boldsymbol{r},t) = du(\boldsymbol{r},t) + p(\boldsymbol{r},t)d\varrho^{-1}(\boldsymbol{r},t) - \sum_{j=1}^{s} \mu_j(\boldsymbol{r},t)dn_j(\boldsymbol{r},t) + + \sum_{j=1}^{s} \tilde{\varphi}_j(\boldsymbol{r},t) \otimes d\tilde{\Phi}_j(\boldsymbol{r},t) , \tag{1.13}$$

where $\tilde{\Phi}_j$ are the fluxes (vectorial, tensorial) and $\tilde{\varphi}_j$ the associated intensive variables, and $\otimes$ indicates a fully contracted tensorial product yielding a scalar. All the coefficients, Θ, p, μ_j, $\tilde{\varphi}_j$, are functionals of the basic variables u, ϱ, n, $\tilde{\Phi}$.

EIT.3) The function η satisfies a balance equation, namely,

$$\frac{\partial}{\partial t}\eta(\boldsymbol{r},t) + \text{div } \boldsymbol{I}_\eta(\boldsymbol{r},t) = \sigma_\eta(\boldsymbol{r},t) , \tag{1.14}$$

with a flux $\boldsymbol{I}_\eta$ and a production σ_η to be determined.

EIT.4) The flux $\boldsymbol{I}_\eta$ is taken as the most general isotropic vector that can be built in terms of the basic macrovariables in Gibbs space.

EIT.5) In some formulations of the theory the condition that $\sigma_\eta(\boldsymbol{r},t) \geq 0$ is imposed for the internal production of EIT entropy.

EIT.6) The entropy production function is constructed in terms of the algebraic invariants associated with the macrovariables in Gibbs space.

EIT.7) The equations of evolution of the Maxwell-Cattaneo-type [Cf. Eq. (1.12)] follow from the preceding items.

EIT.8) Consistently, in the linear flux-free approximation one should recover CIT, as well as the results in equilibrium.

Hence, basically, the theory requires the extension, up to the point deemed necessary, of the space of the state macrovariables, as well as the derivation of their equations of evolution. It should be stressed that the flux of entropy and the entropy production functions associated to the theory, meaning I_η and σ_η of Eq. (1.14), are not given a priori, but they should lead to constitutive equations dependent on the whole set of independent state variables. Maxwell-Cattaneo-Vernotte-type equations for the fluxes are obtained, which yield transport equations for the conserved variables that are of the hyperbolic type and so the problem of instantaneous propagation of thermal perturbation that arises in CIT is avoided. The equations of evolution constitute a set of coupled equations with undetermined kinetic coefficients, to be eventually determined by comparison with experimental data, or derived from a kinetic or statistical theory as described later on this book.

Certainly there are difficulties associated with the construction of EIT; we may mention: (i) How to select and justify the choice of the basic variables; (ii) How — if this is in general possible — to measure such basic variables; (iii) How to interpret the nonequilibrium entropy of Eq. (1.14), that is if it has a well defined physical meaning besides the fact of being an appropriate state function; (iv) We can repeat the question as in (iii), but with reference to the nonequilibrium temperature Θ, as well as the pressure p, chemical potentials μ, and all other thermodynamic variables, in Eq. (1.13); (v) How does EIT compare with other phenomenological nonequilibrium thermodynamics, and if it can be derived from a fundamental kinetic or statistical theory [57, 58].

In general terms we may say that EIT applies from the moment when CIT reaches the limits of its domain of validity. EIT can deal satisfactorily with a wide variety of situations. It is particularly useful in the study of phenomena in the not-so-low-frequencies and not-so-long-wavelengths range, as in sound propagation; diffusion through membranes; light and neutron scattering; propagation of thermal pulses; second sound; etc. Another field of application of EIT is the case of materials with long relaxation times such as occur in polymer solutions, non-Newtonian fluids, dielectrics, glasses and amorphous solids, etc.

As noted, nonequilibrium thermodynamics formulated at a phenomenological level has remained more or less an open subject, particularly in the nonlinear highly dissipative regime outside CIT. In spite of the fact that several efforts, namely those described in RT and EIT, have been made to cover a good deal of such situations, none of them can be considered as fully satisfactory. One of the main reasons for this is that they lack a solid justification based on the general principles of statistical mechanics. The difficulty that arises in this task of deriving irreversible thermodynamics from the microscopic laws of physics resides first of all in the choice of a basic set of variables appropriate to describe the nonequilibrium macroscopic state of the system under consideration. Next to it stands the question of the existence of a functional involving all of these variables which could provide a fundamental relation able to generate the basic relations of the irreversible thermodynamics for such systems. Of course, the well established laws of equilibrium thermodynamics and CIT must be recovered from such fundamental relation as particular limiting cases. Since the entropy function enjoys a preferential status in these cases, one may seek a fundamental relation playing the role of an entropy-like function for the particular descriptions to be used in the study of the macroscopic nonequilibrium states of a many-body system. Finally, to close the theory it is necessary to provide equations for the time evolution of the macroscopic variables.

The underlying ideas behind the questions posed above go back to Maxwell [35] and Boltzmann [62] in the 19th century. Their analyses went beyond equilibrium thermodynamics and constructing evolution equations that have been successfully applied to a large variety of situations from gases to solids [63]. There is certainly a profound physical meaning in Boltzmann equation, but the task of solving the full nonlinear equation and of clarifying its domain of validity has been rather difficult [64, 65]. Nevertheless, very general results are extracted from it such as the general conservation equations for the locally conserved densities and the $\mathcal{H}$-theorem [66]. The relationship between the function $\mathcal{H}$ and the equilibrium entropy of a dilute gas has also been clearly shown [67]. Insofar as its relation with irreversible thermodynamics is concerned, it has been shown that the Chapman-Enskog solution of the Boltzmann equation leads to the laws of CIT, and that the moments solution set forth by Grad [68] leads to one version of EIT [38, 69].

In the hands of Gibbs and Einstein the ideas of Maxwell and Boltzmann were finally brought into a full theory relating the laws of equilibrium thermodynamics to those of microscopic physics. Yet, as some authors have repeatedly emphasized the use of an ensemble to describe the time evolution of a system undergoing irreversible processes has not yet been fully exploited. There are several methods [70–79] that have been successful in describing nonequilibrium phenomena around local equilibrium states of the system, thus providing a basis for CIT and an accompanying fluctuation theory. Others, such as projection-operator methods, first devised by Zwanzig [80, 81] (see also [72, 82, 83]) may be used to deal in principle with situations away from equilibrium, but are of quite difficult manipulation and concrete relations are not easily drawn.

We address next this question of a microscopic approach to irreversible thermodynamics from a statistical-mechanical point of view. This is done by resorting to the so called Nonequilibrium Statistical Operator Method (NESOM) in the context of the formalism of maximization of the informational-statistical entropy (MaxEnt), to be referred to as MaxEnt-NESOM. MaxEnt-NESOM appears to be a formalism founded on sound basis which is concise, practical, and of an appealing structure.

Chapter 2

Nonequilibrium Statistical Operator Method

In this chapter we briefly review the basic tenets of the MaxEnt-NESOM and the construction of the nonequilibrium statistical operator (NESO). In reference [10] its application to the study of dissipative systems with complex behavior was discussed, namely the case of synergetic processes leading to self-organization with the emergence of order out of thermal chaos, the so-called Prigogine dissipative structures [84–87].

As a starting point, we recall that the purpose of Statistical Mechanics of systems away from equilibrium is to determine their thermodynamic properties and the evolution in time of their macroscopic observables, in terms of the dynamical laws which govern the motion of their constitutive elements. As pointed out by Oliver Penrose [88], Statistical Mechanics is notorious for conceptual problems to which it is difficult to give a convincing answer:

- What is the physical significance of a Gibbs ensemble?
- How can we justify the standard ensembles used in equilibrium theory?
- What are the right ensembles for nonequilibrium problems?
- How can we reconcile the reversibility of microscopic mechanics with the irreversibility of macroscopic behavior?

Also, related to the case of systems out of equilibrium, Ryogo Kubo in the oppening address to the Oji Seminar [89] told us that statistical mechanics of nonlinear nonequilibrium phenomena is just in its infancy and further progress can only be hoped by close cooperation with experiment. Some progress has been achieved since then, which is partially described here.

The study of the macroscopic state of nonequilibrium systems presents far greater difficulties than those faced in the theory of equilibrium systems. This is mainly due to the fact that a more detailed discussion is necessary to determine the temporal dependence of measurable properties, and to calculate transport coefficients that are space-dependent and also time-dependent (namely those evolving along with the macrostate of the system in which dissipative processes are unfolding). That dependence is nonlocal in space and time, that is, it encompasses space correlations and memory effects respectively.

It is considered [90] that the basic goals of nonequilibrium statistical mechanics are:

1. to derive transport equations and to grasp their structure;
2. to understand how the approach to equilibrium occurs in natural isolated systems;
3. to study the properties of steady states; and
4. to calculate the instantaneous values and the temporal evolution of the physical quantities which specify the macroscopic state of the system.

Nonequilibrium statistical mechanics has typically followed two directions: One is the kinetic theory of dilute gases, where, starting with a few — albeit controversial — hypotheses, it can be obtained a description of how simple systems evolve and approach equilibrium (the celebrated Boltzmann's transport theory and $\mathcal{H}$-theorem).

An extension of these ideas to dense systems followed several paths like, for example, the construction of a generalized theory of kinetic equations (see for example reference [91, 92]), and the so-called equations of the BBGKY hierarchy (see for example references [93, 94]). The other is the generalization of the theory for Brownian motion, where the complicated dynamic equations — the so called generalized Newton-Langevin equations — which follow from the laws of Mechanics are accompanied by statistical assumptions. To this approach belong for example the formalism of the correlation functions due to Mori (for details see references [95, 96]) and some aspects of the so-called master equation method (for example see reference [94]).

The approaches used to develop a theory encompassing the programme described by items (1) to (4) stated above, have been classified by Zwanzig [90] as:

(a) intuitive techniques;

(b) techniques based on the generalization of the kinetic theory of gases;

(c) techniques based on the theory of stochastic processes;

(d) expansions from an initial equilibrium ensemble; and

(e) generalizations of Gibbs' ensemble algorithm.

This last item (e) is connected with Penrose's question stated above concerning what are the right ensembles for nonequilibrium problems, and with the MaxEnt-NESOM we are considering here. In the absence of a Gibbs-style ensemble approach, for a long time different kinetic theories were used, with variable success, to deal with the great variety of nonequilibrium phenomena occurring in physical systems in nature. A prototype, and a very successful one, has been the famous Boltzmann's transport theory. In connection with it Lebowitz and Montroll has rightly commented [97] that "the beautiful elegance of [Boltzmann] equation, so easily to derive intuitively and so difficult to justify rigorously, is as impressive today as it was over a hundred years ago when it sprang like Minerva fully clothed from the head of Jupiter." The MaxEnt-NESOM seems to provide the grounds, based on sound principles, for a general prescription to choose appropriate ensembles for nonequilibrium systems. The formalism has an accompanying nonlinear quantum kinetic theory of large scope, which encompasses Boltzmann approach as a particular limiting case.

There exist several approaches to the NESOM: some are based on heuristic arguments [70, 98–100], others on projection-operator techniques [72, 83]. It has been shown that these different constructions in NESOM can be unified under a unique variational principle [100, 101]. It consists in the maximization of the Gibbs statistical entropy (sometimes called the fine-grained informational entropy), subjected to certain constraints and including irreversibility, non-locality in space, and memory. NESOM provides a very promising technique that implies in a far-reaching generalization of the statistical methods developed by Boltzmann and Gibbs. It is a formalism that may be considered to be encompassed within the scope of the scheme of Jaynes' Predictive Statistical Mechanics [71, 102, 103], which is a powerful approach based on scientific inference and Bayesian approach in probability [13, 103–105], to be built only on the access to the relevant information that there exists of the system. As pointed out by Jaynes "How shall we best think about Nature and most efficiently predict her behavior, given only our incomplete knowledge [of the microscopic details of the system] [...] we need to see it, not as an example of the N-body equations of motion, but as an example of the logic of scientific inference, which by-passes all details by going directly from our macroscopic information to the best macroscopic predictions that can be made from that information [...] Predictive Statistical Mechanics is not a physical theory, but a method of reasoning that accomplishes this by finding, not the particular that the equations of motion say in any particular case, but the general things that they say in "almost all" cases consistent with our information; for those are the reproducible things."

Again following Jaynes' reasoning, the construction of a statistical approach is based on "a rather basic principle [...]: If any macrophenomenon is found to be reproducible, then it follows that all microscopic details that were not under the experimenters' control must be irrelevant for understanding and predicting it". Further, "the difficulty of prediction from microstates lies [...] in our own lack of the information needed to apply them. We never know the microstate; only a few aspects of the macrostate. Nevertheless, the aforementioned principle of [macroscopic] reproducibility convinces us that this should be enough; the relevant information is there, if only we can see how to recognize it and use it." The technical problem associated to the method is, as put by Jaynes, "how shall we use probability theory to help us do plausible reasoning in situations where, because of incomplete information we cannot use deductive reasoning?" In other words, the main question associated with this approach is how to obtain the probability assignment compatible with the available information while avoiding unwarranted assumptions. This is answered by Jaynes who formulated the criterion that: the least biased probability assignment $\{p_j\}$ for a set of mutually exclusive events $\{x_j\}$ is that which maximizes the quantity S_I, sometimes referred to as the informational entropy, given by

$$S_I = -\sum_j p_j \ln p_j , \tag{2.1}$$

subject to the constraints imposed by the available information [106]. This is based on Shannon's ideas [107], who first demonstrated that, for an exhaustive set of mutually exclusive propositions, there exists a unique function measuring the uncertainty of the probability assignment. This is the variational principle that provides a unifying theoretical framework for the NESOM [64, 65, 101].

The MaxEnt-NESOM appears as a quite appropriate formalism to provide microscopic (statistical-mechanical) foundations to phenomenological irreversible thermodynamics [14, 108, 109], and nonclassical thermo-hydrodynamics well beyond the traditional (classical) approach [110]. MaxEnt-NESOM appears as a very powerful, concise, soundly based, and elegant formalism of, as already noted, broad scope to deal with systems arbitrarily away from equilibrium. Robert Zwanzig wrote that the formalism "has by far the most appealing structure, and may yet become the most effective method for dealing with nonlinear transport processes" [90]. The use of the method in situations in equilibrium and near equilibrium (linear regime) is already a textbook matter [111, 112]. To cover generalized nonlinear nonequilibrium processes in nature, the formalism has been largely developed by several authors: the most appealing approach is that of D. N. Zubarev, initiated in the sixties [100]. It is also presented in other books, as the one by Akhiezer and Peletminskii [113], McLennan [114], and in a number of publications. The formalism is a systematization and a vast generalization of earlier pioneering approaches due to, among others, Kirkwood [115], Green [116], Mori-Oppenheim-Ross [70], Mori [77], Zwanzig [79, 80]. The unification of the approaches is due to Zubarev and Kalashnikov [117], and reviewed in reference [101], on a frame based on the already noted quite engaging line of thought based on information theory and scientific inference set forth by E.T. Jaynes, including nonlocality in space and memory effects (space and time correlations), and nonlinear relaxation effects. The formalism is embedded in the transcendental work of N. N. Bogoliubov [93]. In the context of MaxEnt-NESOM, Robertson derives a nonlinear kinetic theory [72,

118], and further approaches are, among others, due to Zubarev [100], Kalashnikov [119], Peletminskii [99, 113], Vstovskii [120], Sergeev [121], etc., and Lauck et al. derive a generalized nonlinear quantum transport theory [122]. Moreover, elsewhere are described a MaxEnt-NESOM-based response function theory and scattering theory for far-from-equilibrium systems governed by nonlinear kinetic equations, and an accompanying double-time nonequilibrium-thermodynamic Green function formalism [101]. A number of successful applications of the method to the study of experiments in the area of nonlinear transport and nonlinear relaxation effects are presently available in the literature.

The first and fundamental step in MaxEnt-NESOM is the choice of the basic set of variables which are appropriate for the characterization of the macroscopic state of the system. This involves a contracted description (macroscopic or mesoscopic instead of the microscopic of Mechanics) in terms of, say, the dynamical quantities $\hat{P}_1(\boldsymbol{r}), \hat{P}_2(\boldsymbol{r}), \ldots, \hat{P}_n(\boldsymbol{r})$ (Hermitian operators in Quantum Mechanics or functions defined in phase space in Classical Mechanics).

The NESOM statistical operator is a functional of these quantities, to be called $\varrho(\{\hat{P}_j(\boldsymbol{r})\}|t)$, or, for short, $\varrho(t)$. The thermodynamic (macroscopic or mesoscopic) state is characterized by a point in Gibbs space, given at time t by the set of macrovariables $Q_1(\boldsymbol{r},t), Q_2(\boldsymbol{r},t), \ldots, Q_n(\boldsymbol{r},t)$, which are the averages of the $\hat{P}_j$, i.e. $Q_j(\boldsymbol{r},t) = \mathrm{Tr}\{\hat{P}_j(\boldsymbol{r})\, \varrho(t)\}$ (Tr is the usual indication for trace operation). It should be noticed that quantities $\hat{P}_j$ and Q_j depend on position $\boldsymbol{r}$, which is the case when they are local densities, or, eventually, may be homogeneous quantities. The quantities $\hat{P}_j$ change in time with the microscopic evolution of the mechanical state of the system, but in an experiment we do not follow, with the measurement apparatus, this microscopic evolution but the values of the macroscopic variables Q_j. The results of such experiment are to be described by evolution equations of the form

$$\boxed{\frac{\partial}{\partial t} Q_j(\boldsymbol{r},t) = \Phi_j \left\{ Q_j(\boldsymbol{r},t), \ldots, Q_n(\boldsymbol{r},t); \boldsymbol{r}, t \right\} ,} \tag{2.2}$$

where the functionals Φ_j are, in general, nonlinear in the state variables Q_j, nonlocal in space, and carrying memory effects, i.e. depending on the past history of the macroscopic state of the system from time t_0 when the experiment starts up to time t when a measurement is performed.

These considerations immediately rise several questions that need be addressed [90]:

i) How to choose the basic variables? At present there seems to be no wholly satisfactory theory to generate this information to make a unique decision. It has been suggested that this basic set of variables must include all approximate integrals of motion (or quasi-invariant) variables, as it is the case in MaxEnt-NESOM as we shall discuss below.

ii) How are the functionals in Eq. (2.2) obtained? In other words, what is the form of the nonlinear equations of evolution for macrovariables Q_j. There are several approaches available, which are associated to the different techniques corresponding to the items (a) to (e) previously listed in this chapter. In MaxEnt-NESOM the answer is straightforward: once the nonequilibrium statistical operator is given, the equations of evolution are the statistical average over the nonequilibrium ensemble of the mechanical equations of motion of quantities $\hat{P}_j$ (Hamilton equations in the classical case and Heisenberg equations in the quantum case).

iii) The question of the initial conditions. The equations of evolution, Eqs. (2.2), are of first order in the time derivative and therefore they require an initial condition for a unique solution to be obtained. Many times this is done using initial conditions that appear reasonable and well suited to theoretical analysis. The ideal should be to have experimental access to these values, but this is seldom feasible. Thus, in any particular problem the sensitivity of the results to the details of the chosen initial conditions needs be carefully considered.

Finally, it should be noticed that for isolated systems the nonequilibrium statistical operator satisfies Liouville equation, which is reversible in time. This poses another fundamental question, namely,

iv) How to obtain irreversible behavior in the evolution of the macroscopic state of the system? This is sometimes referred to as the time arrow problem (see for example references [123–128]). In MaxEnt-NESOM irreversibility is incorporated from the outset using an *ad hoc* non-mechanical hypothesis, as will be shown later on.

In MaxEnt-NESOM the choice of the basic variables depends on each concrete physical problem under consideration. A guidance is provided by Bogoliubov's procedure of contraction of description [129, 130]. This is based on the existence of a hierarchy of relaxation times, meaning, according to this view, that a succession of contracted descriptions (that is, ever smaller number n of quantities $\hat{P}_j$) is possible if there exists a succession of relaxation times, say $\tau_\mu < \tau_1 < \ldots$, such that after each one has elapsed, correlations with lifetimes smaller that this time length are damped out and can be ignored, and then increasingly shortened basic sets of variables can be used for a proper description of the system macrostate. Generally speaking, the initial state of a nonequilibrium many-body system is determined by as many variables as its number of degrees of freedom. Consider now the particle interaction time, or time of a collision τ_μ, generally referred to as the time for microrandomization. It is roughly given by $r_0/\bar{v}$, where r_0 is the collision length and $\bar{v}$ some average velocity of the particles. After this time has elapsed, a kind of smoothing should occur: details of the initial information are lost, and the macrostate of the system is expected to be specified by fewer variables than in the initial stage. This defines a contracted description, so-called first kinetic stage. The process of reduction of the basic set of variables proceeds in the sequence of time intervals $\tau_\mu, \tau_1, \ldots$, when additional correlations die out, and part of the previously needed variables become dependent on the rest or on a fewer new ones. Such contractions are then determined by the spectrum of relaxation times of the system under consideration. Illustrative examples involving the case of spin relaxation is given in reference [131], and for the case of the highly photoexcited plasma in semiconductors in [132].

Uhlenbeck [130] has pointed out that it seems likely that successive contractions of the description are an essential feature of the theory of irreversible processes, and that, if such contraction is possible, it must be a property of the basic equations of the systems. In a sense this is the case in MaxEnt-NESOM, encompassed in the ideas set forward by, among others, Mori [77], Zubarev [100], and Peletminskii [99]. It consists in introducing a separation of the total Hamiltonian into two parts, namely

$$\boxed{\hat{H} = \hat{H}_0 + \hat{H}' ,} \tag{2.3}$$

where $\hat{H}_0$ is a "relevant" (or secular) part composed of the Hamiltonians of the subsystems involving the kinetic energies and a part of the interactions, namely, those strong enough leading to *very rapid relaxation processes* (meaning those with relaxation times much smaller than the characteristic time scale of the experiment), and possessing certain characteristic properties as shown below. The other term, $\hat{H}'$, contains the interactions related to relaxation processes with *long-time relaxation times.* The said characteristics of the strong interactions depend on the problem under consideration: They require the condition — to be called *Zubarev-Peletminskii law* — that the evolution of the basic dynamical variables with $\hat{H}_0$ produces the linear combinations

$$\boxed{\frac{1}{i\hbar}\left[\hat{P}_j, \hat{H}_0\right] = \sum_{k=1}^{n} \Omega_{jk}\hat{P}_k\,,} \tag{2.4}$$

where $j = 1, 2, \ldots, n$ and $[\hat{P}, \hat{H}_0]$ is the commutator of $\hat{P}$ and $\hat{H}_0$, and in an appropriate representation the $\Omega's$ are c-numbers determined by $\hat{H}_0$. As previously noticed quantities $\hat{P}$ can be dependent on the space variable, that is when considering local densities of dynamical quantities, and then quantities Ω can also be differential operators.

Zubarev-Peletminskii law of Eq. (2.4) provides a selection rule for the choice of the basic set of variables: First, the secular part of the Hamiltonian, viz. $\hat{H}_0$, has to be adequately chosen in each particular problem under consideration: as noted, it contains the kinetic energies plus the interactions strong enough to produce damping of correlations in times smaller than that of the characteristic time of the experiment — typically the experimental resolution time; hence, at this point Bogoliubov's principle of contraction of description is at work. Second, one introduces a few dynamical variables $\hat{P}$ deemed relevant for the description of the physical problem in hands, and next the commutator of them with $\hat{H}_0$ is calculated. The dynamical variables — different from those already introduced — that appear on the linear combination indicated by the right-hand side of Eq. (2.4) are incorporated to the basic set. This procedure is then repeated until a closure is obtained. As discussed in next chapter, the procedure is relevant for the choice of the basic variables in phenomenological irreversible thermodynamics [133]. Practical use of the method requires to introduce an appropriate truncation procedure along the chain that the application of the method produces [134, 135]. In this case a careful evaluation is required concerning the amount of information lost with such truncation, and if its order of magnitude is negligible as compared to the one kept as relevant. It needs be stressed that Eq. (2.4) also encompasses the case of quantities $\hat{P}$ which have associated null coefficients Ω, i.e. they are constants of motion under the dynamics generated by $\hat{H}_0$. Accordingly they are acceptable basic variables, and $\hat{H}_0$ itself falls under this condition. Consequently, Peletminskii-Zubarev selection rule implies, taking into account all dynamical quantities, that, under the dynamics generated by $\hat{H}_0$, are kept in the subspace of the Hilbert state space that they span, and then are referred to as quasi-conserved variables. It needs be noticed that the procedure is, in a sense, the analog of the choice of the basic variables in the case of systems in equilibrium, when it is done on the basis of taking the wholly conserved ones.

Assuming that the basic set $\{\hat{P}_j\}$ has been chosen, the NESO is built in MaxEnt-NESOM, within the context of Jaynes' Predictive Statistical Mechanics, resorting to the principle of maximization of the informational-statistical entropy (MaxEnt) with memory and *ad hoc* hypotheses which introduce from the outset irreversible evolution from an initial condition of preparation of the system. The procedure creates a mimic of Prigogine's dynamical condition for dissipativity, an approach to introduce a time arrow in dynamics as described in references [125, 136]. Details are given in references [64, 65, 101], which we very briefly review next in order to clarify its main aspects.

The equivalent of Eq. (2.1) for a physical many-body system is Gibbs' statistical entropy (or fine-grained informational-statistical entropy)

$$S_G(t) = -\,\mathrm{Tr}\{\varrho(t)\ln\varrho(t)\}\,, \tag{2.5}$$

with the NESO $\varrho(t)$ normalized at all times in the interval (t_0, t), namely,

$$\mathrm{Tr}\{\varrho(t')\} = 1\,, \tag{2.6}$$

for $t_0 \le t' \le t$, where t_0 is the time of initial preparation of the system on which is performed a measurement at time t.

Following MaxEnt we obtain the best choice for the NESO $\varrho(t)$ looking for a maximum of $S_G(t)$ under the constraint of Eq. (2.6) and the conditions that in $t_0 \le t' \le t$,

$$Q_j(\mathbf{r}, t') = \mathrm{Tr}\{\hat{P}_j(\mathbf{r})\varrho(t')\}\,, \tag{2.7}$$

where $j = 1, 2, \ldots, n$, being implicit that the choice of the set of basic variables $\{\hat{P}_j\}$ has been performed according to the procedure already discussed. The set of Eqs. (2.7) introduces a dynamical character in the chosen information, but it needs be noticed that the information-gathering interval (t_0, t) can (and should) be reduced to information recorded at a unique time: This is because the macroscopic state of the system is characterized by the set of basic variables $\{Q_j(\mathbf{r}, t)\}$, which satisfy the equations of evolution — in the kinetic theory that the formalism produces — which have a well defined solution once their values are given at a unique time (besides of course the boundary spatial conditions).

This means that the MaxEnt-NESOM yields information on the macrostate of the system at time t, when a measurement is performed, including the evolutionary history (in the interval from the initial time of preparation t_0 up to time t) by which the system came into that state (what introduces a generalization of Kirkwood's time-smoothing formalism [115]).

Following well known procedures to solve the variational problem with constraints, namely the method of Lagrange multipliers, it is obtained that [101]

$$\varrho(t) = \exp\left\{ -\psi(t) - \sum_{j=1}^{n} \int d^3r \int_{t_0}^{t} dt' \varphi_j(\mathbf{r};t,t')\hat{P}_j(\mathbf{r},t'-t) \right\}, \tag{2.8}$$

which has the form of a generalized Gibbsian canonical distribution, and the inclusion of memory effects is clearly evidenced. In this Eq. (2.8), ψ is the Lagrange multiplier which ensures the normalization of $\varrho(t)$, that is

$$\psi(t) = \ln \mathrm{Tr}\left\{ \exp\left[- \sum_{j=1}^{n} \int d^3r \int_{t_0}^{t} dt' \varphi_j(\mathbf{r};t,t')\hat{P}_j(\mathbf{r},t'-t) \right] \right\}, \tag{2.9}$$

and φ_j are Lagrange multipliers associated to the constraints of Eqs. (2.7) and quantities $\hat{P}_j$ are given in Heisenberg representation. Furthermore, ψ is a functional of these φ_j and can be interpreted as the logarithm of a generalized nonequilibrium partition function.

Next it is introduced an extra assumption on the form of the Lagrange multipliers. This is done in order to have a theory which, first, generates irreversible behavior in the evolution of the macroscopic state of the system, second, introduces a set of variables $\{F_j(\mathbf{r},t)\}$ that play the role of intensive variables thermodynamically conjugated to the extensive macrovariables $\{Q_j(\mathbf{r},t)\}$, in such a way to generate a complete connection with phenomenological irreversible thermodynamics (to be shown in next chapter), and, third, fixes an initial condition from which the irreversible evolution of the macrostate of the system proceeds. This is accomplished with the definition

$$\varphi_j(\boldsymbol{r};t,t') = w(t,t')F_j(\boldsymbol{r},t') , \tag{2.10}$$

where $w(t,t')$ is an auxiliary weight function with well defined properties described and discussed in detail elsewhere [64, 65, 101]. Functions $w(t,t')$ are typically — at least in the existing approaches to NESOM — kernels that appear in the mathematical theory of convergence of integrals. At this point we choose a particular one, which seemingly, on the one side, appears as the most appropriate for calculations, and on the main side, for providing the best physical picture of the final results: this is Abel's kernel proposed by Zubarev's in his approach to MaxEnt-NESOM [100], namely

$$w(t,t') = \varepsilon \exp\{\varepsilon(t'-t)\} , \tag{2.11}$$

and it is assumed preparation of the system in the remote past, $t_0 \to -\infty$, meaning to neglect fast transients between the initial preparation and the time when the first measurement is performed. Moreover, ε is a positive infinitesimal that goes to zero after the calculation of traces in the determination of average values has been performed.

Using the definitions given by Eq. (2.10) and (2.11), it follows that $\varrho(t)$ of Eq. (2.8) takes the form [64, 65, 100, 101]

$$\begin{aligned} \varrho_\varepsilon(t) &= \exp\left\{\varepsilon \int_{-\infty}^{t} dt' e^{\varepsilon(t'-t)} \ln \bar{\varrho}(t',t'-t)\right\} = \\ &= \exp\left\{\ln \bar{\varrho}(t,0) - \int_{-\infty}^{t} dt' e^{\varepsilon(t'-t)} \frac{d}{dt} \ln \bar{\varrho}(t',t'-t)\right\} , \end{aligned} \tag{2.12}$$

where the last term follows after partial integration in time, and it has been introduced the auxiliary NESO $\bar{\varrho}$, which plays an important role in the theory. This auxiliary distribution (sometimes refered-to as the coarse grained part of the NESO, or a "frozen" quasi-equilibrium distribution)) is given by

$$\bar{\varrho}(t,0) = \exp\left\{-\phi(t) - \sum_{j=1}^{n} \int d^3r F_j(\boldsymbol{r},t)\hat{P}_j(\boldsymbol{r})\right\} , \tag{2.13}$$

with

$$\boxed{\phi(t) = \ln \mathrm{Tr}\left\{\exp\left[-\sum_{j=1}^{n}\int d^3r F_j(\boldsymbol{r},t)\hat{P}_j(\boldsymbol{r})\right]\right\},} \tag{2.14}$$

and

$$\bar{\varrho}(t',t'-t) = \exp\left\{-\frac{1}{i\hbar}(t'-t)\hat{H}\right\}\bar{\varrho}(t',0)\exp\left\{\frac{1}{i\hbar}(t'-t)\hat{H}\right\}. \tag{2.15}$$

From Eq. (2.12) it follows that the initial condition is

$$\boxed{\varrho_\varepsilon(t_0) = \bar{\varrho}(t_0,0)\,,} \tag{2.16}$$

what amounts to an initial description (preparation) of the system neglecting all correlations among the basic variables prior to time t_0, what amounts to a kind of generalized Stosszahlansatz at the initial time. Furthermore, using Eq. (2.12) it can be shown that the NESO can be separated into two parts, namely

$$\boxed{\varrho_\varepsilon(t) = \bar{\varrho}(t,0) + \varrho'_\varepsilon(t)\,,} \tag{2.17}$$

and a time-dependent projection operator $\mathcal{P}_\varepsilon(t)$ can be defined, which has the property that

$$\mathcal{P}_\varepsilon(t)\ln\varrho_\varepsilon(t) = \ln\bar{\varrho}(t,0)\,. \tag{2.18}$$

Thus $\mathcal{P}_\varepsilon$ projects the logarithm of the NESO over the logarithm of the auxiliary distribution [64, 65, 101].

The quantity $\hat{S}(t,0) = -\ln\bar{\varrho}(t,0) = -\mathcal{P}_\varepsilon(t)\ln\varrho_\varepsilon(t)$ is called the informational entropy operator. Its average over the nonequilibrium ensemble is the informational entropy to be defined in next chapter [cf. Eq (3.2) below]. The characteristic and properties of this operator are described in [137]. Let us next discuss these results.

The procedure, we recall, involves the macroscopic description of the system in terms of a set of macrovariables $\{Q_j(\boldsymbol{r},t)\}$, which are the average over the nonequilibrium ensemble — characterized by the NESO $\varrho_\varepsilon(t)$ — of a set of dynamical quantities $\{\hat{P}_j(\boldsymbol{r})\}$. The latter are quasi-conserved quantities, in the sense that under the dynamics generated by $\hat{H}_0$ they satisfy Eq. (2.4). They are called "relevant" variables, and we denote the space they define as the informational space. The remaining quantities for the dynamical description of the system, namely, those absent from the informational space associated to the constraints in MaxEnt [Cf. Eqs. (2.7)], are called the "irrelevant" variables. In term of these considerations, the role of the projection operator of Eq. (2.18) can be better characterized: It introduces what is referred to as a *coarse-graining procedure* when it projects the logarithm of $\varrho_\varepsilon(t)$ onto the subspace of the relevant (or informational) variables, this projection being the logarithm of the auxiliary NESO, namely $\ln\bar{\varrho}(t,0)$. This is graphically illustrated in Fig. 2.1 for the average values of these quantities. Hence, the procedure eliminates the "irrelevant" variables, given *an expression depending only on the chosen set of informationally relevant variables.* The "irrelevant" variables are hidden in the NESO, as given by Eq. (2.17), in the contribution $\varrho'_\varepsilon(t)$ since it is a functional of the last term in the exponential in Eq. (2.12), where the time derivative drives it outside the subspace of relevant variables. The coarse-graining performed by the projection operator is time dependent, implying that the operation is dependent on the present, at time t, macroscopic state of the system described by the $Q_j(\boldsymbol{r},t)$. In the next chapter further relevance of this coarse-graining projection will become better clarified when related to the definitions of MaxEnt entropy and entropy production and their connection with irreversible thermodynamics.

Two further comments are worth introducing, whose details are given in references [64, 65, 101]. First, for a given dynamical quantity $\hat{A}$, its average value in MaxEnt-NESOM (i.e. the value to be compared with an experimental one) is given by

$$\langle\hat{A}|t\rangle = \lim_{\varepsilon\to 0}\mathrm{Tr}\{\hat{A}\varrho_\varepsilon(t)\} = $$

$$= \lim_{\varepsilon\to 0}\mathrm{Tr}\left\{\hat{A}\exp\left[\varepsilon\int_{-\infty}^{t} dt' e^{\varepsilon(t'-t)}\ln\bar{\varrho}(t',t'-t)\right]\right\}. \tag{2.19}$$

In Eq. (2.19) the limit of ε going to zero is taken, we recall, after the regular average (the trace operation) has been performed. This operation introduces in the formalism the so-called Bogoliubov's method of quasi-averages [138]. Bogoliubov's procedure involves a symmetry-breaking process which is introduced in order to remove degeneracies connected with one or several groups of transformations in the description of the

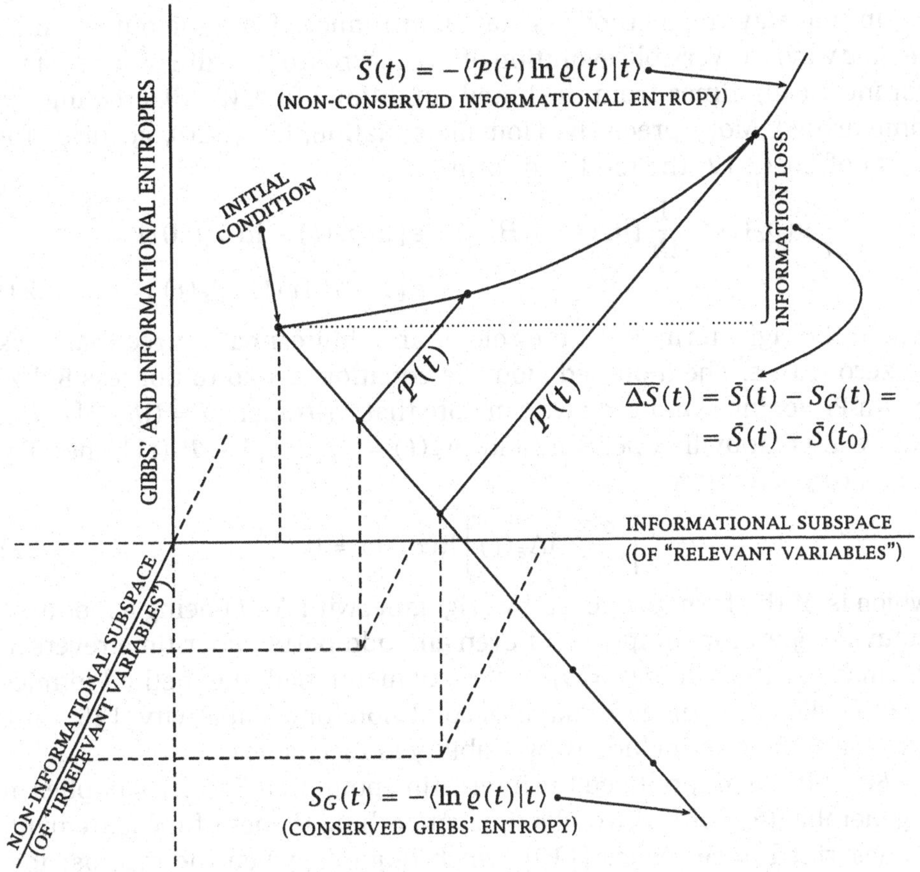

Figure 2.1: An outline of the description of the non-equilibrium-dissipative macroscopic state of the system. The projection — depending on the instantaneous macrostate of the system — introduces the coarse-graining procedure consisting into the projection onto the subspace of the "relevant" variables associated to the informational constraints in MaxEnt-NESOM.

system. In the present case *the symmetry-breaking is that of the time-reversal symmetry*. The presence of Abel's kernel $\varepsilon \exp\{\varepsilon(t' - t)\}$, with its given properties, selects the subset of retarded solutions from the total group of solutions of Liouville equation

$$\frac{\partial}{\partial t} \ln \varrho(t) + \frac{1}{i\hbar} \left[\ln \varrho(t), \hat{H} \right] = 0 \,. \tag{2.20}$$

In that way the method, as stated, introduces from the outset (in an *ad hoc* way) irreversible evolution of the macroscopic state of the system for increasing times from the initial condition of Eq. (2.16) (creating the time arrow). More precisely, Liouville equation, Eq. (2.20), acquires for $\varrho_\varepsilon(t)$ of Eq. (2.12) the modified form

$$\frac{\partial}{\partial t}\ln\varrho_\varepsilon(t) + \frac{1}{i\hbar}[\ln\varrho_\varepsilon(t), H] = -\varepsilon[\ln\varrho_\varepsilon(t) - \ln\bar{\varrho}(t,0)] = \\ = -\varepsilon[1 - \mathcal{P}_\varepsilon(t)]\ln\varrho_\varepsilon(t) , \qquad (2.21)$$

where the right term plays the role of an infinitesimal source that goes to zero with ε. The modified Liouville equation is said to possess Boltzmann-Prigogine symmetry: we can note that writting $i\mathcal{L}\varrho = (i\hbar)^{-1}[\varrho, \hat{H}]$, where $\mathcal{L}$ is Liouville operator, and $i\Lambda_\varepsilon(t) = i\mathcal{L} + \varepsilon[1 - \mathcal{P}_\varepsilon(t)]$, then Eq. (2.21) can be written as

$$\left[\frac{\partial}{\partial t} + i\Lambda_\varepsilon(t)\right]\ln\varrho_\varepsilon(t) = 0 , \qquad (2.22)$$

which is of the form proposed by Prigogine, with $\Lambda_\varepsilon(t)$ being a modified Liouville operator composed of even and odd parts under time-reversal. We may say that, in that way, as already mentioned, the method mimics the so-called Prigogine's dynamical condition for dissipativity [125, 136], which attempts to include irreversibility in mechanics.

Second, we may noticed the interesting point that Eq. (2.19) implies in a generalization of Kirkwood's time-smoothing theory of measurement, as described in reference [115], which requires, when the macroscopic state is changing with time, the identification of a macroscopic quantity through a time-smoothing-like procedure as in Eq. (2.19). It implies a two-step operation in which the statistical average from an initial distribution (at time t_0) is followed by the weighted time average up to the time a measurement is performed.

Finally, to close the formalism and to be able to perform calculations of average values of observables, of response functions, and of thermodynamic functions, it is necessary to obtain the equations of evolution for the basic variables. They are given by

$$\frac{\partial}{\partial t}Q_j(\mathbf{r}, t) = \mathrm{Tr}\left\{\frac{1}{i\hbar}\left[\hat{P}_j(\mathbf{r}), \hat{H}\right]\varrho_\varepsilon(t)\right\} . \qquad (2.23)$$

Equations (2.23), with $j = 1, 2, \ldots, n$, are Heisenberg's mechanical equations of motion averaged over the nonequilibrium ensemble, which constitute a set of coupled highly nonlinear equations which, in general, requires an extremely difficult mathematical handling. We recall that ε goes to zero after the trace operation has been performed. They are of the form of Eqs. (2.2); in fact, $\varrho_\varepsilon(t)$ of Eq. (2.12) is a functional of the set of variables $\{F_j(\boldsymbol{r},t)\}$, which are related to the set of variables $\{Q_j(\boldsymbol{r},t)\}$ by Eqs. (2.7). We note — for the sake of completeness but without going into details — that it can be shown [64, 65, 101, 122] that Eq. (2.23) admits the alternative form

$$\boxed{\frac{\partial}{\partial t}Q_j(\boldsymbol{r},t) = J_j^{(0)}(\boldsymbol{r},t) + J_j^{(1)}(\boldsymbol{r},t) + \mathcal{J}_j(\boldsymbol{r},t)\,,} \tag{2.24}$$

where

$$J_j^{(0)}(\boldsymbol{r},t) = \mathrm{Tr}\left\{\frac{1}{i\hbar}\left[\hat{P}_j(\boldsymbol{r}),\hat{H}_0\right]\bar{\varrho}(t,0)\right\}\,, \tag{2.25a}$$

$$J_j^{(1)}(\boldsymbol{r},t) = \mathrm{Tr}\left\{\frac{1}{i\hbar}\left[\hat{P}_j(\boldsymbol{r}),\hat{H}'\right]\bar{\varrho}(t,0)\right\}\,, \tag{2.25b}$$

$$\mathcal{J}_j(\boldsymbol{r},t) = \mathrm{Tr}\left\{\frac{1}{i\hbar}\left[\hat{P}_j(\boldsymbol{r}),\hat{H}'\right]\varrho'_\varepsilon(t)\right\}\,, \tag{2.25c}$$

with $\bar{\varrho}$ and ϱ'_ε defined in Eq. (2.17), and $\hat{H}_0$ and $\hat{H}'$ in Eq. (2.3).

The collision operator $\mathcal{J}_j$ is extremely difficult to handle. However, it can be rewritten, through the use of the closure condition of Eq. (2.4) and the separation of the NESO as given by Eq. (2.17) (the technical details are given in reference [122]) in the form of an infinite series of terms in the form

$$\mathcal{J}_j(\boldsymbol{r},t) = \sum_{l=2}^{\infty} J_j^{(l)}(\boldsymbol{r},t)\,. \tag{2.26}$$

In this equation $J_j^{(l)}$, with $l = 2,3,\ldots$, are partial collision operators which are instantaneous in time and organized in increasing power l in the interaction strengths contained in $\hat{H}'$. They are of an ever increasing level of complexity with increasing l. It is worth noticing that these contributions are composed of several terms, consisting of (1) the mechanical effects of collisions (l-particle collisions), (2) terms that account for the evolution of the thermodynamic state of the system while the collision processes take place, and (3) terms arising out of memory effects.

The results here summarized allow us to state that the MaxEnt-NESOM provides a generalized transport theory of a far-reaching scope, which incorporates nonlinearity, nonlocality in space and time (space correlations and memory effects respectively), and may cover quite arbitrary situations of nonequilibrium in many-body systems. We draw the attention to the relevant fact that the generalized time-smoothing procedure (in Kirkwood's sense) weighted by the kernel of Eq. (2.11), is reflected in the final equations of evolution as a kind of fading memory [the advanced solutions are discarded; cf. Eq. (2.21)]. In that way, the macroscopic state of the system evolves irreversibly towards the future. Moreover, if after an initial nonequilibrium state of preparation the system is left free of any exciting source, and kept in contact with ideal reservoirs, say, of energy and particles, the macrostate of the system evolves towards a final state of equilibrium described by the usual grand-canonical distribution.

Once the general overview of the MaxEnt-NESOM has been presented, we proceed in next chapter to briefly describe the connection of the last two chapters in providing a framework for Informational Statistical Thermodynamics (IST).

Chapter 3

Informational Statistical Thermodynamics

The term IST is applied to the theory that provides a statistical-mechanical-founded irreversible thermodynamics on the basis of Jaynes' Predictive Statistical Mechanics with the accompanying maximum entropy formalism (MaxEnt). The use of the underlying ideas of MaxEnt to derive macroscopic properties for systems in nonequilibrium were pursued by several authors. After Jaynes' original papers were published [106] there followed applications in which the general features of irreversibility were discussed [30, 139]. Kinetic and transport equations using MaxEnt were obtained for very specific cases [140, 141], afterwards brought together into a unifying scheme by Lewis [142] (See also references [143, 144]). Connections with CIT are due to several authors, in particular to Zubarev [145]. Finally, we should mention Nettleton's use of Maxent to obtain a description of conduction of heat in dense fluids [146, 147], and a generalized Grad-type foundation for EIT with altered thermodynamic forces [40, 41]. Nettleton pursued this line in further papers [42, 148, 149]. A version of IST founded on MaxEnt-NESOM, that is, a microscopic approach to phenomenological irreversible thermodynamics based on such particular nonequilibrium ensemble formalism is described in references [133, 134]. It provides, as a particular case, the connection of EIT of the first chapter and the NESOM of chapter 2 [14].

To build the theory the first step is the choice of the basic variables. This task, performed heuristically in EIT, contrasts with the MaxEnt-NESOM where it follows from the use of Eqs. (2.3) and (2.4) as described in chapter 2: Once the Hamiltonian is separated in $\hat{H}_0$ and $\hat{H}'$ (on the basis of the existence of a distinct hierarchy of time scales in Bogoliubov's sense [93, 129, 130]) one proceeds with the application of Zubarev-Peletminskii law of Eq. (2.4), which provides a selection rule for the choice of the basic set of variables. As already noted, the closure procedure may not follow in a finite number of steps and then an appropriate truncation procedure needs be introduced [14, 135]. A difficulty arises with the introduction of the truncation procedure. In fact, there is not, in principle, a well defined expansion parameter — as, for example, provided by the Knudsen number in the Chapman-Enskog method of solution of Boltzmann equation — and so there is not a clear indication about the order of magnitude of the neglected terms. This appears, at first sight, as a shortcoming of the method, similar to the one stressed by van Kampen [150] in the case of Grad's moment method. The classical Grad's moments method in kinetic theory, proposed over forty years ago, has been the subject of deep interest. It provided a mesoscopic background to irreversible thermodynamics: Just as the Chapman-Enskog method of solving the Boltzmann equation was used by Prigogine [31] to derive the basics of CIT, Grad's moment method has been taken to provide justification for some efforts to extend the scope of validity of CIT, in particular EIT.

However, there are indications that IST contains in itself a way to overcome the just considered question of evaluation of the domain of validity of the truncation procedure [135]. The question has been considered in the particular case of a photogenerated plasma in semiconductors [134, 151, 152] where a criterion for the order of the truncation procedure was introduced. Furthermore, concerning the question of the role of fast-relaxing contributions in $\hat{H}_0$, an illustrative example can be found in the analysis of the terms arising in the dissipative evolution of the carrier system in the photoinjected plasma in semiconductors, the fastest one being the result of the strong and long-range Coulomb interaction (leading to relaxation times in the femtosecond time scale; see for example refs. [132, 151–155]).

Once the choice of the basic variables has been performed, the auxiliary (or coarse-grained) NESO of Eq. (2.13) is constructed, and the nonequilibrium distribution ϱ_ε follows from Eq. (2.12). The dynamical quantities $\hat{P}_j$ are scalars, vectors, or tensors depending on the case, and so are the macrovariables Q_j, and the Lagrange multipliers F_j. The connection with irreversible thermodynamics, that is, the construction of IST, is done through the definition of a state function, namely the MaxEnt-NESOM-based expression for the statistical entropy.

Gibbs entropy, the straightforward generalization of equilibrium and the one used in the variational approach to the MaxEnt-NESOM, namely

$$S_G(t) = -\operatorname{Tr}\{\varrho_\varepsilon(t) \ln \varrho_\varepsilon(t)\} , \tag{3.1}$$

cannot represent a, say, appropriate quasi-thermodynamic entropy since it is a constant of the motion, that is $dS_G(t)/dt = 0$. (This is the consequence of the fact that ϱ_ε satisfies a modified Liouville equation, namely Eq. (2.21), with the presence of sources that, we recall, vanish after the trace operation in the calculation of averages has been performed). This is a manifestation of the fact that S_G is a fine-grained entropy that preserves information. This is the information provided at the initial time of preparation of the system as indicated by Eq. (2.16), the one given in terms of the initial values of the set of basic variables $\{Q_j(\mathbf{r}, t_0)\}$. Hence, for any subsequent time $t \geq t_0$ it is introduced a coarse-grained MaxEnt-NESOM (or informational-statistical) entropy such that, according to the foundations of the formalism, is associated to the information provided by the constraints of Eqs. (2.7) at each time $t \geq t_0$. In that way, the difference between the entropy to be defined, $\bar{S}(t)$, and Gibbs entropy of Eq. (3.1) is a kind of measurement of the information lost when the macroscopic state of the system is described in terms of the reduced set of basic variables and only them, namely in terms of what we have called the informational subspace composed of the "relevant" variables. Hence, on that basis and with the help of Fig. 2.1, it is defined the coarse-grained MaxEnt-NESOM entropy

$$\boxed{\bar{S}(t) = -\operatorname{Tr}\{\varrho_\varepsilon(t)\mathcal{P}_\varepsilon(t) \ln \varrho_\varepsilon(t)\} = -\operatorname{Tr}\{\varrho_\varepsilon(t) \ln \bar{\varrho}(t,0)\} ,} \tag{3.2}$$

which is given by the average value of the projection of the logarithm of the fine-grained NESO [Cf. Eq. (2.18)] over the subspace containing the basic set of dynamical quantities, that is, the informational subspace. This MaxEnt-NESOM entropy (or informational-statistical entropy) is to be placed in correspondence with those of phenomenological theories. From the definition of Eq. (3.2) it is possible to retrieve, as particular cases, the corresponding expressions in equilibrium, in CIT, and in EIT. Hence, the IST built on its basis [14, 108, 109] can be considered a broad generalization of existing phenomenological approaches. Let us consider this point further.

Using the expression for the auxiliary (coarse-grained) $\bar{\varrho}(t,0)$ as given by Eq. (2.13), we can write the MaxEnt-NESOM entropy of Eq. (3.2) as

$$\boxed{\bar{S}(t) = \int d^3r\, \bar{\eta}(\boldsymbol{r},t) = \phi(t) + \int d^3r \sum_{j=1}^{n} F_j(\boldsymbol{r},t) Q_j(\boldsymbol{r},t)\,,} \tag{3.3}$$

where $\bar{\eta}(\boldsymbol{r},t)$ represents the local density of informational-statistical entropy [1]

Moreover, the differential of the MaxEnt-NESOM entropy, at space position $\boldsymbol{r}$ and time t, satisfies the Pffafian-like form (generalized Gibbs relation)

$$d\bar{\eta}(\boldsymbol{r},t) = \sum_{j=1}^{n} F_j(\boldsymbol{r},t) dQ_j(\boldsymbol{r},t)\,, \tag{3.4}$$

corresponding to a generalization of the expression in EIT of Eq. (1.13).

The equations of evolution, Eqs. (2.2), for the basic variables, defined in Eqs. (2.23) or (2.24), as given by the formalism, are, in general, balance equations of the form

$$\frac{\partial}{\partial t} Q_j(\boldsymbol{r},t) = -\operatorname{div} \boldsymbol{I}_j(\boldsymbol{r},t) + \xi_j(\boldsymbol{r},t)\,, \tag{3.5}$$

where $\boldsymbol{I}_j$ is the flux of variable Q_j, and ξ_j accounts for sources and/or sinks of such variable [This is a simplified notation; let us recall that the Q_j can be scalars, vectors and tensors, that is, scalar densities, their vectorial fluxes, and higher order (tensorial) fluxes] [156, 157]. Using Eqs. (3.5) we can write a balance equation for the MaxEnt-NESOM entropy density, namely

$$\frac{\partial}{\partial t} \bar{\eta}(\boldsymbol{r},t) + \operatorname{div} \boldsymbol{I}_\eta(\boldsymbol{r},t) = \sigma_\eta(\boldsymbol{r},t)\,, \tag{3.6}$$

where $\boldsymbol{I}_\eta$ is the flux of MaxEnt-NESOM entropy defined as

$$\boldsymbol{I}_\eta(\boldsymbol{r},t) = \sum_{j=1}^{n} F_j(\boldsymbol{r},t) \boldsymbol{I}_j(\boldsymbol{r},t), \tag{3.7}$$

[1] Comment: We call the attention to the fact that in classical hydrodynamics it is defined an entropy $s(\boldsymbol{r},t)$ per density of mass $\varrho(\boldsymbol{r},t)$, the equivalence then being $\bar{\eta} \equiv \varrho(\boldsymbol{r},t) s(\boldsymbol{r},t)$ (see [94]).

and σ_η involves sources and/or sinks in the form

$$\sigma_\eta(\boldsymbol{r},t) = \sum_{j=1}^{n} \left\{ \boldsymbol{I}_j(\boldsymbol{r},t) \cdot \nabla F_j(\boldsymbol{r},t) + F_j(\boldsymbol{r},t)\xi_j(\boldsymbol{r},t) \right\} . \tag{3.8}$$

Equation (3.6) is the general form that the balance entropy equation has in IST. Although it resembles the one postulated in EIT, it is far more general since the $\boldsymbol{I}_\eta$ is not necessarily restricted to the general physical fluxes of hydrodynamics nor any statement is a priori made about the sign of σ. Thus the correspondence between the phenomenological and statistical approaches to irreversible thermodynamics has been established. Also, from Eq. (3.4) we can see that

$$\boxed{F_j(\boldsymbol{r},t) = \delta\bar{S}(t)/\delta Q_j(\boldsymbol{r},t) ,} \tag{3.9}$$

where δ stands for functional differential. This Eq. (3.9) defines the Lagrange parameters as the differential coefficients of the informational-statistical entropy. The set of Eqs. (3.9) relates the intensive macrovariables F_j to the basic macrovariables, Q_j (on which $\bar{S}$ depends), and may be considered as *equations of state for nonequilibrium systems.*

The Maxwell-Cattaneo-Vernotte-type equations of EIT, as is the case of Eq. (1.12), follow from the equations of evolution, Eq. (2.24), in a completely generalized way. This point deserves additional comments. In CIT, the fluxes are related to the thermodynamic forces through phenomenological linear relations, v.g. Fourier's (flux of energy proportional to the gradient of temperatures) and Fick's (flux of matter proportional to the gradient of concentration) laws [Cf. Eqs. (1.9)]. As already noted in chapter 1, this leads to parabolic-type equations of diffusion [Cf. Eqs. (1.10)], with the uncomfortable question of propagation of thermal or diffusive perturbations with infinite velocity and difficulties when theoretical results are compared with the experimental data at high frequencies and short wavelengths. Once the fluxes are elevated to the rank of basic variables (as proposed in EIT), the phenomenological equations of CIT are replaced by equations of evolution of the MCV-type. The equations of diffusion of CIT are replaced by equations of the telegraphist type, hyperbolic partial differential equations implying wave propagation with finite velocity and damping. Furthermore, IST provides these equations with explicit introduction of the microscopic dynamics that governs the movement and collision processes in the system of particles. Details are given elsewhere, consisting in deriving the equations of evolution (MCV-type) in condensed matter systems, in a local in space and time approximation and linear in the fluxes [133, 134, 154, 158]. This task is being presently extended to non-local conditions [159, 160], in a non-local in time (memory dependent) approach [161], and nonlinear in the fluxes [162]. Transport coefficients are determined at the microscopic (statistical-mechanical) level, in particular mobility and diffusion coefficients and a generalization of Einstein relation between these coefficients in a nonlinear charge transport domain [152, 163, 164]. Also it is derived a generalized hydrodynamics at classical [110] and quantum [165] levels. Resorting to a partial hydrodynamical approach the propagation of thermal waves has been considered in [155, 158].

We recall that in phenomenological EIT, one of the basic postulates is that of a local and instantaneous definite-positive entropy production [25, 56]. In IST the entropy production follows from time derivation of Eq. (3.3) [cf. Eq. (3.4)] to obtain that [166]

$$\boxed{\bar{\sigma}(t) = \frac{d}{dt}\bar{S}(t) = \int d^3r \frac{\partial \bar{\eta}(\mathbf{r},t)}{\partial t}}\,, \tag{3.10}$$

where

$$\frac{d\bar{\eta}(\mathbf{r},t)}{dt} \equiv \bar{\sigma}(\mathbf{r},t) = \sum_{j=1}^{n} F_j(\mathbf{r},t)\frac{\partial}{\partial t}Q_j(\mathbf{r},t)\,, \tag{3.11}$$

with the time derivative following from Eq. (2.24). No definitive assertion concerning the sign of $\bar{\sigma}(\mathbf{r},t)$ can be made in general, due to its extremely complicated expression. However, it is possible to prove that [166] (see Section 4.8)

$$\bar{S}(t) - \bar{S}(t_0) = \bar{S}(t) - S_G(t) \geq 0\,. \tag{3.12}$$

Equation (3.12), with the help of Eq. (3.10), can be alternatively written as

$$\boxed{\int_{t_0}^{t} dt' \int d^3r\, \bar{\sigma}(\mathbf{r},t) \geq 0}\,, \tag{3.13}$$

(t_0 as before is the initial time of preparation of the system), which is an expression stating that the MaxEnt-IST entropy cannot decrease in time.

We call Eq. (3.13) a *weak principle of non-negative informational-entropy production*. This result is equivalent to that obtained by del Rio and García-Colín [167], who interpret the resulting inequality as the fact that a sequence of observations performed (within a precisely specified time interval) on a macroscopic system undergoing irreversible processes, always results in a loss of information in Shannon-Brillouin's [107, 168] sense. This is illustrated in Fig. 2.1: The difference between both kinds of entropies (the fine-grained and the coarse-grained), namely the average of $[1 - \mathcal{P}_\varepsilon(t)] \ln \varrho_\varepsilon(t)$ represents the increase in informational-statistical entropy along the irreversible evolution of the system [Cf. Eqs. (3.12) and (3.2)].

Boltzmann equation and $\mathcal{H}$-theorem are retrieved as particular limiting cases of the complete IST theory [64, 65]. Also, it can be shown that IST, based on the MaxEnt-NESOM, provides statistical-mechanical generalizations to several theorems in generalized phenomenological thermodynamics [101, 166]. One is Prigogine's minimum entropy production law [17, 20, 169], already described in chapter 1, which, we recall, establishes a variational principle consisting in the fact that in the linear regime around a steady state of the system in near equilibrium conditions, the rate of change in time of the entropy production due to internal relaxation processes is negative. In a sense it is a principle of economy: the system accomodates itself in a state that expends entropy as least as possible. Other is a generalization of Glansdorff-Prigogine thermodynamic criterion for evolution [20, 169]: according to it the evolution of the system in any nonequilibrium condition proceeds in such a way to make negative the rate of variation of the entropy production due exclusively to the change in time of the entropy differential coefficients, i.e. the variables F_j in Eq. (3.9), that is [cf. Eq. (3.11)] the quantity (see Section 4.9)

$$\frac{d_F}{dt}\sigma(t) = \sum_{j=1}^{n} \int d^3r \, \frac{\partial}{\partial t} F_j(\mathbf{r}, t) \frac{\partial}{\partial t} Q_j(\mathbf{r}, t) \leq 0 \,. \tag{3.14}$$

Equation (3.14) contains the generalization of the minimum entropy production theorem in the limit of the linear regime around equilibrium. It can be noticed that this criterion for evolution is not a variational principle since it does not involve an exact differential, as it is the case in the linear regime near equilibrium when the entropy production associated to internal relaxation processes in the steady state can play a role of a thermodynamic potential. It seems not to be possible to define appropriate thermodynamic potentials to describe the state of arbitrary nonequilibrium systems, differently to the case of equilibrium and the already mentioned linear regime which allow such approach. An application of the MaxEnt-NESOM to the study of the thermodynamics of a simplified model of a nonequilibrium semiconductor, which illustrates the two principles just described, is given in reference [170].

Finally, a generalization of Glansdorff-Prigogine (in)stability criterion [20, 169] can be derived. According to this criterion the steady state of an open system sufficiently far-from-equilibrium becomes unstable when the so-called excess entropy production function, namely the difference between the internal entropy production in a state slightly deviated from the steady state with that at the steady state, changes sign. This is the physical expression of the otherwise mathematical method of linear stability analysis. For the instability to occur the macroscopic state of the system — governed by nonlinear equations of evolution — must be sufficiently away from equilibrium, since in the linear regime around equilibrium the minimum entropy production theorem ensures its stability. Hence, in the nonlinear regime, in order for instabilities to occur a thermodynamic potential, with an associated variational principle, needs not exist, as already noted above. The stability of the linear regime is a consequence of the inevitable regression of fluctuations, and always there follows dissipation of order. On the other hand, in the nonlinear regime, at the point of instability fluctuations can be enhanced and stabilized in a new state. This is the result of a "tug of war" between dissipative effects and positive feedback effects originating in the nonlinear terms in the equations of evolution. The point of instability, characterized by a critical value of the intensity of external forces, defines what is called a bifurcation point of the solutions of the coupled set of equations of evolution, in the sense that at that point a new and stable solution emerges from the, now unstable, previous solution (see for example reference [171]). There follows a transition between two steady states, with the new stable state displaying some kind of long range order on the macroscopic scale, or self-organization, comprising Prigogine's *dissipative structures* already referred to in the introduction as related to theory of dynamical systems with *complex behavior* [84–87, 169, 172]. In this way arises the connection of MaxEnt-NESOM-based IST with the question of dissipation, order, and complexity in natural systems.

Furthermore, we draw attention to the fact of the difficulties involved with the definitions of nonequilibrium entropy, and nonequilibrium temperature (as well as nonequilibrium chemical potential, , pressure, etc.). In the case of the entropy, whether it can be a proper state function depends on a proper choice of the basic macrovariables having been made. As we have seen this is a soluble problem in equilibrium and local equilibrium conditions, but not in the general case since there is not a wholly satisfactory theory to proceed to a unique and well defined choice.

The question of the definition and measurement of a nonequilibrium temperature is a quite interesting subject. A quantity that can be defined in EIT and IST, by analogy with equilibrium, and dubbed quasitemperature, $\Theta(\boldsymbol{r}, t)$, has a reciprocal given by

$$\Theta_l^{-1}(\boldsymbol{r}, t) = k_B \delta \bar{S}(t) / \delta \varepsilon_l(\boldsymbol{r}, t), \tag{3.15}$$

where $\varepsilon_l(\boldsymbol{r}, t)$ is the macrovariable energy density, and the index l means individual definitions for, eventually, different subsystems of the given material. It is relevant to keep in mind that, opposite to situations of equilibrium and in CIT when the temperature depends only on energy and particle number, in IST Θ depends on all the basic macrovariables, including the fluxes. This question has been considered within the phenomenological framework of EIT, and a possible, albeit controverted, experiment for its characterization proposed [173]; the case of a photoinjected plasma in semiconductors under a constant electric field is considered in [174]. Certainly, an important point is: Can the nonequilibrium temperature of Eq. (3.15) be measured? This should be so in order to have a consistent and acceptable theory, since it is a basic intensive macrovariable (a Lagrange multiplier in IST) associated with the macroscopic description of the system. It is shown [101, 175] that response functions of the system in an experiment are dependent on the basic intensive variables and, therefore, the experiment can provide a way to measure quasitemperature as defined by Eq. (3.15). Detailed considerations and comparison with experiment are presented elsewhere [176].

Without going into further details, summarizing we may say that the MaxEnt-NESOM (based on Jaynes' Predictive Statistical Mechanics and related to Shannon-Brillouin's theory of information) is an adequate formalism to provide, in the scheme of IST, microscopic (statistical-mechanical) foundations to irreversible thermodynamics.

Chapter 4

PROPERTIES OF THE INFORMATIONAL STATISTICAL THERMODYNAMIC ENTROPY

In this Chapter we further analize the informational-statistical entropy in IST which we have defined in the previous Chapter, namely $\bar{S}(t)$ of Eq. (3.2) [see also Eqs. (3.3) and (3.4)].

For that purpose we find it most convenient to use a specific — and general — nonequilibrium ensemble formalism, which consists in a nonequilibrium generalized grand-canonical ensemble [156], however for systems describable in terms of single-particle states.

4.1 Nonequilibrium Grand-Canonical Ensemble

Consider a system of many bosons or fermions, with a single-particle field operator

$$\psi_\sigma(\boldsymbol{r}) = \sum_{\boldsymbol{k}} \varphi_{\boldsymbol{k}\sigma}(\boldsymbol{r}) c_{\boldsymbol{k}\sigma} \tag{4.1}$$

where φ is a single-particle wavefunction in a state characterized by orbital wavevector $\boldsymbol{k}$ and spin index σ (in the particular case of quantized lattice vibrations, that is, of phonons, the index σ is to be interpreted as the index for the corresponding branch, namely transverse or longitudinal vibrations of acoustic or optical type), and $c(c^\dagger)$ are, as usual, annihiliation (creation) operators in the corresponding states. For fermions and bosons are satisfied the known (anti)commutation relations, and let $\varepsilon_{k\sigma}$ be the corresponding single-particle energies, with the eigenfunctions being the $\varphi_{k\sigma}$ above). We introduce the single-particle density operator

$$\hat{n}(\boldsymbol{r}) = \sum_\sigma \psi_\sigma^\dagger(\boldsymbol{r})\psi_\sigma(\boldsymbol{r}) \,, \tag{4.2a}$$

and the single-particle energy density operator

$$\hat{h}(\boldsymbol{r}) = \sum_\sigma \psi_\sigma^\dagger(\boldsymbol{r})\hat{\varepsilon}_\sigma(\boldsymbol{r})\psi_\sigma(\boldsymbol{r}) \,, \tag{4.2b}$$

where $\hat{\varepsilon}$ is the single-particle Hamiltonian operator. But, practical reasons make it convenient to introduce their $\boldsymbol{Q}$-wavevector Fourier transforms (that is, to work in reciprocal space), namely

$$\hat{n}(\boldsymbol{Q}) = \int d^3r\, e^{i\boldsymbol{Q}\cdot\boldsymbol{r}}\hat{n}(\boldsymbol{r}) = \sum_{\mathbf{k}\sigma} \hat{n}_{\boldsymbol{kQ}\sigma} \,, \tag{4.3a}$$

$$\hat{h}(\boldsymbol{Q}) = \int d^3r e^{i\boldsymbol{Q}\cdot\boldsymbol{r}}\hat{h}(\boldsymbol{r}) = \sum_{\mathbf{k}\sigma} E_{\boldsymbol{kQ}\sigma}\hat{n}_{\boldsymbol{kQ}\sigma} \,, \tag{4.3b}$$

where

$$E_{\boldsymbol{kQ}\sigma} = \frac{1}{2}(\varepsilon_{\boldsymbol{k}+\frac{1}{2}\boldsymbol{Q},\sigma} + \varepsilon_{\boldsymbol{k}-\frac{1}{2}\boldsymbol{Q},\sigma}) \,, \tag{4.4a}$$

and

$$\hat{n}_{\boldsymbol{kQ}\sigma} = c^\dagger_{\boldsymbol{k}+\frac{1}{2}\boldsymbol{Q},\sigma} c^\dagger_{\boldsymbol{k}-\frac{1}{2}\boldsymbol{Q},\sigma} \tag{4.4b}$$

is Dirac-Landau-Wigner single-particle density dynamical operator. Moreover, for simplicity — but without losing generality — we have taken plane wave states for the φ's of Eq. (4.1); the volume of the system is taken equal to 1.

We write for the system Hamiltonian, taken in the form given by Eq. (2.3),

$$\hat{H}_0 = \sum_{k\sigma} \varepsilon_{k\sigma}c^\dagger_{k\sigma}c_{k\sigma} + \hat{H}_\Sigma + \hat{H}_R \,, \tag{4.5}$$

$$\hat{H}' = \hat{H}'_{S\Sigma} + \hat{H}'_{SR} \ , \tag{4.6}$$

where the first term on the right of Eq. (4.5) is the Hamiltonian of the free particles of the system, while $\hat{H}_\Sigma$ and $\hat{H}_R$ are, respectively, the Hamiltonians of the free sources and reservoirs to which the system is coupled. In Eq. (4.6), the two terms on the right account for the interactions between the system and sources and reservoirs. Explicit expressions of $\hat{H}_\Sigma$, $\hat{H}_R$, $\hat{H}'_{S\Sigma}$, and $\hat{H}'_{SR}$ are not necessary for our purposes next.

We recall once again that, for simplicity, we are considering many-particle systems which admit a description in a single-quasiparticle picture. This is typically the case of solid state matter, and within it we may highlight the very important area of semiconductor physics, which, besides the associated scientific interest, is of large relevance in those technological and industrial aspects associated to the development of electronic and optoelectronic devices. Moreover, and this is of fundamental importance, there exists a vast number of experimental studies in these materials, mainly concerning transport and optical properties, which consist in measurements of extremely high quality and with very high levels of resolution. In this case the single-particles are for the electron system Landau's quasi-electrons in Bloch band states and impurity states, and as excitons, and displaying collective excitations like plasmons and magnons once Coulomb interaction is dealt with in the random-phase approximation; for the lattice system are the phonons in the different branches of vibration and polarization; and finally the different kind of hybrid excitations like polarons, polaritons, plasmaritons, phonoritons, and so on.

We proceed to build in the framework of MaxEnt-NESOM, and for the system described by the Hamiltonians of Eqs. (4.5) and (4.6), a corresponding nonequilibrium grand-canonical ensemble. For that purpose we start with a description which introduces, to begin with, the operators $\hat{n}$ and $\hat{h}$ of Eqs. (4.2), or more precisely, with those of Eqs. (4.3). We call the attention to the fact that the contributions $\boldsymbol{Q} = \boldsymbol{0}$ are the global ones, that is,

$$\hat{n}(\boldsymbol{Q} = \boldsymbol{0}) = \int dr^3 \hat{n}(\boldsymbol{r}) = \hat{N} \ , \tag{4.7a}$$

$$\hat{h}(\boldsymbol{Q} = \boldsymbol{0}) = \int dr^3 \hat{h}(\boldsymbol{r}) = \sum_{\boldsymbol{k}\sigma} \varepsilon_{\boldsymbol{k}\sigma} c^\dagger_{\boldsymbol{k}\sigma} c_{\boldsymbol{k}\sigma} \ , \tag{4.7b}$$

i.e., the particle number and free part of the Hamiltonian of the system (the volume, we recall, is 1).

Next, we need to look for the verification of the selection rule (Zubarev-Peletminskii symmetry condition) of Eq. (2.4), which, as shown below, introduces in the picture the fluxes of all order of the two basic densities. First we calculate

$$\frac{1}{i\hbar}[\hat{n}(\boldsymbol{Q}),\hat{H}_0] = \mathrm{i}\sum_{\boldsymbol{k}\sigma} \Delta\omega_{\boldsymbol{kQ}\sigma}\hat{n}_{\boldsymbol{kQ}\sigma}\,, \tag{4.8a}$$

$$\frac{1}{i\hbar}[\hat{h}(\boldsymbol{Q}),\hat{H}_0] = \mathrm{i}\sum_{\boldsymbol{k}\sigma} E_{\boldsymbol{kQ}\sigma}\Delta\omega_{\boldsymbol{kQ}\sigma}\hat{n}_{\boldsymbol{kQ}\sigma}\,, \tag{4.8b}$$

where

$$\Delta\omega_{\boldsymbol{kQ}\sigma} = \omega_{\boldsymbol{k}+\frac{1}{2}\boldsymbol{Q},\sigma} - \omega_{\boldsymbol{k}-\frac{1}{2}\boldsymbol{Q},\sigma}\,; \qquad \hbar\omega_{\boldsymbol{k}\sigma} = \varepsilon_{\boldsymbol{k}\sigma}\,, \tag{4.9}$$

with $\Delta\omega_{\boldsymbol{kQ}\sigma}$ indicating Bohr's frequency for a transition between states $\boldsymbol{k} - \boldsymbol{Q}/2$ and $\boldsymbol{k} + \boldsymbol{Q}/2$. Let us consider this quantity: An expansion in powers of $\boldsymbol{Q}$ of Bohr's frequency of Eq. (4.9) is given by

$$\Delta\omega_{\boldsymbol{kQ}\sigma} = \boldsymbol{Q}\cdot\Big\{\nabla_{\boldsymbol{k}}\omega_{\boldsymbol{k}\sigma} + \frac{1}{2!}\nabla_{\boldsymbol{k}}(\boldsymbol{Q}\cdot\nabla_{\boldsymbol{k}}\omega_{\boldsymbol{k}\sigma}) + \frac{1}{3!}\nabla_{\boldsymbol{k}}[\boldsymbol{Q}\cdot\nabla_{\boldsymbol{k}}(\boldsymbol{Q}\cdot\nabla_{\boldsymbol{k}}\omega_{\boldsymbol{k}\sigma})] + \cdots\Big\}\,, \tag{4.10}$$

where $\nabla_{\mathbf{k}}$ is the gradient in $\boldsymbol{k}$-space. Next, using the vector relation

$$\nabla(\boldsymbol{A}\cdot\boldsymbol{B}) = \boldsymbol{A}\times(\nabla\times\boldsymbol{B}) + \boldsymbol{B}\times(\nabla\times\boldsymbol{A}) + (\boldsymbol{B}\cdot\nabla)\boldsymbol{A} + (\boldsymbol{A}\cdot\nabla)\boldsymbol{B}\,, \tag{4.11}$$

where $\times$ stands for vectorial product, it can be proved that

$$\nabla_{\boldsymbol{k}}(\boldsymbol{Q}\cdot\nabla_{\boldsymbol{k}}\omega_{\boldsymbol{k},\sigma}) = (\boldsymbol{Q}\cdot\nabla_{\boldsymbol{k}})\nabla_{\boldsymbol{k}}\omega_{\boldsymbol{k},\sigma}\,, \tag{4.12}$$

$$\nabla_{\boldsymbol{k}}[\boldsymbol{Q}\cdot\nabla_{\boldsymbol{k}}(\boldsymbol{Q}\cdot\nabla_{\boldsymbol{k}}\omega_{\boldsymbol{k}})] = (\boldsymbol{Q}\cdot\nabla_{\boldsymbol{k}})^2\nabla_{\boldsymbol{k}}\omega_{\boldsymbol{k}}\,, \tag{4.13}$$

and so on, and after some algebra one finally obtains that

$$\Delta\omega_{\boldsymbol{kQ}\sigma} = \omega_{\boldsymbol{k}+\frac{1}{2}\boldsymbol{Q},\sigma} - \omega_{\boldsymbol{k}-\frac{1}{2}\boldsymbol{Q},\sigma} = \boldsymbol{Q}\cdot\boldsymbol{u}(\boldsymbol{k},\boldsymbol{Q},\sigma)\,, \tag{4.14}$$

with $\boldsymbol{u}$ given by

$$\boldsymbol{u}(\boldsymbol{k},\boldsymbol{Q},\sigma) = \sum_{l=0}^{\infty}\frac{1}{(2l+1)!}\Big(\frac{\boldsymbol{Q}}{2}\cdot\nabla_{\boldsymbol{k}}\Big)^{2l}\nabla_{\boldsymbol{k}}\omega_{\boldsymbol{k}\sigma}\,, \tag{4.15}$$

which has a very peculiar form, namely, the first contribution ($l = 0$) is the group velocity of the single-particle in state $|\boldsymbol{k}\sigma\rangle$, followed by the curvature of this $\boldsymbol{k}$-dependent group velocity, etc. It may be noticed that the series involves an expansion in ever increasing even powers of Q, the wavenumber which is the reciprocal of the wavelength $2\pi/Q$, and, therefore, one may expect the first terms to be the leading ones in processes involving excitations with only the longest wavelengths (implying in the near quasi-homogeneous situations of the classical hydrodynamic limit). Hence, Eq. (4.8a) can be rewritten as

$$\frac{1}{i\hbar}\left[\hat{n}(\boldsymbol{Q}), \hat{H}_0\right] = i\boldsymbol{Q} \cdot \sum_{\mathbf{k}} \boldsymbol{u}(\boldsymbol{k}, \boldsymbol{Q}, \sigma)\hat{n}_{\boldsymbol{k}\boldsymbol{Q}\sigma} , \tag{4.16}$$

from which we define the quantity

$$\hat{\boldsymbol{I}}_n(\boldsymbol{Q}) = \sum_{\mathbf{k}} \boldsymbol{u}(\boldsymbol{k}, \boldsymbol{Q}, \sigma)\hat{n}_{\boldsymbol{k}\boldsymbol{Q}\sigma} , \tag{4.17}$$

present in Eq. (4.16), as the flux of particle number. We recall that the quantity $\hat{n}_{\boldsymbol{k}\boldsymbol{Q}\sigma}$ is the Dirac-Landau-Wigner single-particle dynamical density operator of Eq. (4.4b).

Proceeding in a similar way we find for the energy density operator that [cf. Eq. (4.8b)]

$$\frac{1}{i\hbar}\left[\hat{h}(\boldsymbol{Q}), \hat{H}_0\right] = i\boldsymbol{Q} \cdot \hat{\boldsymbol{I}}_h(\boldsymbol{Q}) , \tag{4.18}$$

where

$$\hat{\boldsymbol{I}}_h(\boldsymbol{Q}) = \sum_{\boldsymbol{k}\sigma} E_{\boldsymbol{k}\boldsymbol{Q}\sigma}\boldsymbol{u}(\boldsymbol{k}, \boldsymbol{Q}, \sigma)\hat{n}_{\boldsymbol{k}\boldsymbol{Q}\sigma} . \tag{4.19}$$

Transforming back to the direct space the Heisenberg equations of motion for the quantities $\hat{n}$ and $\hat{h}$, defined in Eq. (4.2a) and (4.2b), are

$$\frac{\partial}{\partial t}\hat{n}(\boldsymbol{r}) = \frac{1}{i\hbar}\left[\hat{n}(\boldsymbol{r}), \hat{H}_0 + \hat{H}'\right], \tag{4.20a}$$

$$\frac{\partial}{\partial t}\hat{h}(\boldsymbol{r}) = \frac{1}{i\hbar}\left[\hat{h}(\boldsymbol{r}), \hat{H}_0 + \hat{H}'\right], \tag{4.20b}$$

once we use the Hamiltonian of Eq. (2.3) together with Eqs. (4.5) and (4.6). Using the results provided by Eqs. (4.16), (4.17), (4.18), and (4.19) we find that

$$\frac{\partial}{\partial t}\hat{n}(\boldsymbol{r}) + \nabla \cdot \hat{\boldsymbol{I}}_n(\boldsymbol{r}) = \frac{1}{i\hbar}\left[\hat{n}(\boldsymbol{r}), \hat{H}'\right] , \tag{4.21a}$$

$$\frac{\partial}{\partial t}\hat{h}(\boldsymbol{r}) + \nabla \cdot \hat{\boldsymbol{I}}_h(\boldsymbol{r}) = \frac{1}{i\hbar}\left[\hat{h}(\boldsymbol{r}), \hat{H}'\right] , \tag{4.21b}$$

where $\nabla \cdot$ is the divergence operator, and $\hat{\boldsymbol{I}}_n$ and $\hat{\boldsymbol{I}}_h$ are to be interpreted as the first order fluxes (or vectorial fluxes) of single-particles' density and energy density and the right-hand side accounts for the collision processes governed by $\hat{H}'$. According to the closure condition both fluxes are incorporated as basic variables. Next, the rule of Eq. (2.4) is applied to them, to obtain

$$\frac{1}{i\hbar}\left[\hat{\boldsymbol{I}}_n(\boldsymbol{Q}), \hat{H}_0\right] = i\boldsymbol{Q} \cdot \hat{I}_n^{[2]}(\boldsymbol{Q}) , \tag{4.22a}$$

$$\frac{1}{i\hbar}\left[\hat{\boldsymbol{I}}_h(\boldsymbol{Q}), \hat{H}_0\right] = i\boldsymbol{Q} \cdot \hat{I}_h^{[2]}(\boldsymbol{Q}) , \tag{4.22b}$$

where

$$\hat{I}_n^{[2]}(\boldsymbol{Q}) = \sum_{\boldsymbol{k}\sigma} [\boldsymbol{u}(\boldsymbol{k}, \boldsymbol{Q}, \sigma)\boldsymbol{u}(\boldsymbol{k}, \boldsymbol{Q}, \sigma)]\, \hat{n}_{\boldsymbol{k}\boldsymbol{Q}\sigma} , \tag{4.23a}$$

$$\hat{I}_h^{[2]}(\boldsymbol{Q}) = \sum_{\boldsymbol{k}\sigma} E_{\boldsymbol{k}\boldsymbol{Q}\sigma} [\boldsymbol{u}(\boldsymbol{k}, \boldsymbol{Q}, \sigma)\boldsymbol{u}(\boldsymbol{k}, \boldsymbol{Q}, \sigma)]\, \hat{n}_{\boldsymbol{k}\boldsymbol{Q}\sigma} , \tag{4.23b}$$

are second rank tensors (the term $[\boldsymbol{u}\,\boldsymbol{u}]$ between square brackets stands for tensorial product of vectors [177]) which constitute the second order fluxes (or flux of the flux) of density and energy density operators, respectively. Therefore, they are to be included in the basic set, and so on, that is, the selecting rule commands that the fluxes of all order of the two basic densities must be considered as basic dynamical variables. The basic set is then composed by

$$\{\hat{h}(\boldsymbol{Q}), \hat{n}(\boldsymbol{Q}), \hat{\boldsymbol{I}}_h(\boldsymbol{Q}), \hat{\boldsymbol{I}}_n(\boldsymbol{Q}), \{\hat{I}_h^{[r]}(\boldsymbol{Q})\}, \{\hat{I}_n^{[r]}(\boldsymbol{Q})\}\} , \tag{4.24}$$

where $r \geq 2$ indicates the order (and also tensorial rank, with $r = 1$ being the vectorial ones) of the fluxes, given by

$$\hat{I}_n^{[r]}(\boldsymbol{Q}) = \sum_{\boldsymbol{k}\sigma} u^{[r]}(\boldsymbol{k}, \boldsymbol{Q}, \sigma)\hat{n}_{\boldsymbol{k}\boldsymbol{Q}\sigma} , \tag{4.25a}$$

$$\hat{I}_h^{[r]}(\boldsymbol{Q}) = \sum_{\boldsymbol{k}\sigma} E_{\boldsymbol{k}\boldsymbol{Q}\sigma} u^{[r]}(\boldsymbol{k}, \boldsymbol{Q}, \sigma)\hat{n}_{\boldsymbol{k}\boldsymbol{Q}\sigma} , \tag{4.25b}$$

where

$$u^{[r]}(\boldsymbol{k}, \boldsymbol{Q}, \sigma) = [\boldsymbol{u}(\boldsymbol{k}, \boldsymbol{Q}, \sigma) \ldots (r \text{ times}) \ldots \boldsymbol{u}(\boldsymbol{k}, \boldsymbol{Q}, \sigma)] , \tag{4.25c}$$

indicating tensorial product of r vectors $\boldsymbol{u}$, producing a tensor of rank r [177].

Equation (4.24) provides the set of basic variables that the Zubarev-Peletminskii selection rule in MaxEnt-NESOM introduces for this problem in thermo-hydrodynamics; it can be noticed that the procedure of Eq. (2.4) introduces a closure condition for the conserving part of the equations of evolution (the left side of Eqs. (4.21)), and similarly for the fluxes, namely

$$\frac{\partial}{\partial t}\hat{I}_n^{[r]}(\boldsymbol{r}) + \nabla \cdot \hat{I}_n^{[r+1]}(\boldsymbol{r}) = \frac{1}{i\hbar}[\hat{I}_n^{[r]}(\boldsymbol{r}), \hat{H}'] \,, \tag{4.26a}$$

$$\frac{\partial}{\partial t}\hat{I}_R^{[r]}(\boldsymbol{r}) + \nabla \cdot \hat{I}_h^{[r+1]}(\boldsymbol{r}) = \frac{1}{i\hbar}[\hat{I}_R^{[r]}(\boldsymbol{r}), \hat{H}'] \,. \tag{4.26b}$$

The construction of the nonequilibrium grand-canonical ensemble follow from the fact that given the basic set of dynamical variables of Eq. (4.24), the nonequilibrium statistical operator $\varrho_\epsilon(t)$ is built in MaxEnt-NESOM [cf. Eq. (2.12)] in terms of the auxiliary NESO $\bar{\varrho}$ of Eq. (2.13). The auxiliary nonequilibrium grand-canonical statistical operator for a closed system composed of the system of interest (to be simply called the system in what follows), sources, and reservoirs, once we take sources and reservoirs as ideal, that is, neglecting the changes in their states by the action of the coupling with the system, can be written as a direct product of the auxiliary statistical distribution of the system, $\bar{\varrho}_S(t,0)$, with the steady-state statistical distributions of the free sources and reservoirs, ϱ_Σ and ϱ_R. The auxiliary distribution of the system is given in this case by

$$\begin{aligned}\bar{\varrho}_S(t,0) = \exp\Big\{&-\phi(t) - \sum_{Q}\Big[F_h(Q,t)\hat{h}(Q) + F_n(Q,t)\hat{n}(Q)\Big] \\ &- \sum_{Q}\Big[\boldsymbol{\alpha}_h(Q,t)\cdot\hat{\boldsymbol{I}}_h(Q) + \boldsymbol{\alpha}_n(Q,t)\cdot\hat{\boldsymbol{I}}_n(Q)\Big] \\ &- \sum_{Q}\sum_{r\geq 2}\Big[\alpha_h^{[r]}(Q,t)\otimes\hat{I}_h^{[r]}(Q) + \alpha_n^{[r]}(Q,t)\otimes\hat{I}_n^{[r]}(Q)\Big]\Big\} \,,\end{aligned} \tag{4.27}$$

once we take into account that the basic set of dynamical variables is given by Eq. (4.24), and where $\otimes$ stands for fully contracted product of tensors, and we have introduced the corresponding set of associated

Lagrange multipliers indicated by

$$\{F_h(Q,t), F_n(Q,t), \boldsymbol{\alpha}_h(Q,t), \boldsymbol{\alpha}_n(Q,t), \{\alpha_h^{[r]}(Q,t)\}, \{\alpha_n^{[r]}(Q,t)\}\}. \tag{4.28}$$

As already noticed, these Lagrange multipliers constitute a set of intensive variables, that, alternatively, completely describe the nonequilibrium thermodynamic state of the system. They are related to the basic set of macrovariables by the relations (which are the equivalent of equations of state in arbitrary nonequilibrium conditions)

$$n(Q,t) = \mathrm{Tr}\left\{\hat{n}(Q)\bar{\varrho}(t,0)\right\} = -\frac{\delta\phi(t)}{\delta F_n(Q,t)}, \tag{4.29a}$$

$$h(Q,t) = \mathrm{Tr}\left\{\hat{h}(Q)\bar{\varrho}(t,0)\right\} = -\frac{\delta\phi(t)}{\delta F_h(Q,t)}, \tag{4.29b}$$

etc., where we have used that ϱ_ϵ and $\bar{\varrho}$, at each time t, define the same average values for the set of basic variables of Eq. (4.24) and only for these [64, 65, 100, 101, 117], and δ stands for functional differential [178]. In these equations, as noticed, we have that $\bar{\varrho}(t,0) = \bar{\varrho}_S(t,0)\varrho_\Sigma\, \varrho_R$, but the distributions of the sources and reservoirs do not contribute in the calculation of the traces once they are, evidently, defined in different state spaces than that of the system. Hence, the basic set of macrovariables is given by

$$\{h(Q,t), n(Q,t), \boldsymbol{I}_h(Q,t), \boldsymbol{I}_n(Q,t), \{I_h^{[r]}(Q,t)\}, \{I_n^{[r]}(Q,t)\}\}, \tag{4.30}$$

which define the nonequilibrium thermodynamic space of the states in IST (in this case we can refer to it as the Gibbs' grand-canonical space of macrostates).

We stress that the true nonequilibrium grand-canonical statistical operator is given in terms of the auxiliary one $\bar{\varrho}(t,0) = \bar{\varrho}_S(t,0)\varrho_\Sigma\varrho_R$ by Eq. (2.12), that is

$$\varrho_\varepsilon(t) = \exp\Big\{ln\,[\bar{\varrho}_S(t,0)\varrho_\Sigma\varrho_R] - \int_{-\infty}^{t} dt'\, e^{\varepsilon(t'-t)}\frac{d}{dt'}\ln[\bar{\varrho}_S(t',t'-t)\varrho_\Sigma(t'-t)\varrho_R(t'-t)]\Big\},$$

with the time evolution as indicated by Eq. (2.15).

4.2 Kinetic theory in terms of the fluxes: A nonclassical hydrodynamics

This nonequilibrium grand-canonical operator describes a fluid of particles and provides foundation for a nonclassical hydrodynamics. In fact, the equations of evolution, Eq. (2.23) or (2.24), for the basic variables of Eq. (4.30), are given by

$$\frac{\partial}{\partial t}h(\mathbf{r},t) + \nabla \cdot \mathbf{I}_h(\mathbf{r},t) = J_h^{(1)}(\mathbf{r},t) + \mathcal{J}_h(\mathbf{r},t)\,, \tag{4.31a}$$

$$\frac{\partial}{\partial t}n(\mathbf{r},t) + \nabla \cdot \mathbf{I}_n(\mathbf{r},t) = J_n^{(1)}(\mathbf{r},t) + \mathcal{J}_n(\mathbf{r},t)\,, \tag{4.31b}$$

$$\frac{\partial}{\partial t}\mathbf{I}_h(\mathbf{r},t) + \nabla \cdot I_h^{[2]}(\mathbf{r},t) = \mathbf{J}_h^{(1)}(\mathbf{r},t) + \mathbf{J}_h(\mathbf{r},t)\,, \tag{4.31c}$$

$$\frac{\partial}{\partial t}\mathbf{I}_n(\mathbf{r},t) + \nabla \cdot I_n^{[2]}(\mathbf{r},t) = \mathbf{J}_n^{(1)}(\mathbf{r},t) + \mathbf{J}_n(\mathbf{r},t)\,, \tag{4.31d}$$

$$\frac{\partial}{\partial t}I_h^{[r]}(\mathbf{r},t) + \nabla \cdot I_h^{[r+1]}(\mathbf{r},t) = J_h^{(1)[r]}(\mathbf{r},t) + \mathcal{J}_h^{[r]}(\mathbf{r},t)\,, \tag{4.31e}$$

$$\frac{\partial}{\partial t}I_n^{[r]}(\mathbf{r},t) + \nabla \cdot I_n^{[r+1]}(\mathbf{r},t) = J_n^{(1)[r]}(\mathbf{r},t) + \mathcal{J}_n^{[r]}(\mathbf{r},t)\,, \tag{4.31f}$$

with $r \geq 2$, where

$$J_h^{(1)}(\mathbf{r},t) = \mathrm{Tr}\left\{\frac{1}{i\hbar}\left[\hat{h}(\mathbf{r}),\hat{H}'\right]\bar{\varrho}(t,0)\right\}, \tag{4.32a}$$

$$\mathcal{J}_h(\mathbf{r},t) = \mathrm{Tr}\left\{\frac{1}{i\hbar}\left[\hat{h}(\mathbf{r}),\hat{H}'\right]\varrho'_\epsilon(t)\right\}, \tag{4.32b}$$

and similarly for all the other quantities. We recall that $\bar{\varrho}(t,0) = \bar{\varrho}_S(t,0)\varrho_\Sigma\varrho_R$, and that $\varrho'_\epsilon(t)$ is the difference between the statistical operator, $\varrho_\varepsilon(t)$, and the auxiliary one, $\bar{\varrho}(t,0)$, as given by Eq. (2.17).

It can be noticed that if the right-hand side of Eqs. (4.31) is neglected we obtain the conservation equation for each quantity, namely for the densities and their fluxes of all order. Moreover, we recall that the terms with the presence of the divergence operator have their origin in $\hat{H}_0$ [they are the terms $J_j^{(0)}$ in Eq. (2.25a)], and that these terms, as well as those of the type $J_j^{(1)}$ in that equations, do not contribute to the production of informational-entropy, what is shown in Appendix II. Hence, they are not associated with dissipative effects, which are exclusively dependent on the collision operators $\mathcal{J}_j$ which are determined by $\hat{H}'$ of Eq. (2.3) and

$\varrho'_\epsilon(t)$ of Eq. (2.17). Hence, for $\hat{H}' = 0$ (a "collisionless system") when also ϱ'_ϵ is null, no dissipation is present.

Equations (4.31) provide the thermo-hydrodynamics of the nonequilibrium many-body system, which, clearly, constitute an infinite set of coupled equations. Evidently, practical usage requires to introduce a truncation procedure, meaning the analog of the one in the Hilbert-Chapmann-Enskog approach to Boltzmann equation. This is discussed in reference [135], and a particular case — a photoinjected plasma in semiconductors — is considered in [133]. In section 4.5 [see p. 70] we present a simple example for partial illustration. Previously, we present a brief discussion of the Lagrange multipliers and thermodynamic fluxes.

4.3 The Lagrange multipliers

Let us consider the Lagrange multipliers (intensive thermodynamic variables) of Eq. (4.28). Those associated to energy and particle densities will be discussed in particular in next subsection; we concentrate here on those associate to the fluxes. As already noted they are determined by the average values of the specific macrovariables in the basic set that defines the nonequilibrium thermodynamic state of the system, and they constitute an alternative closed set of (intensive) nonequilibrium thermodynamic variables that also provide that description. Explicitly,

$$n(\boldsymbol{r},t) = \mathrm{Tr}\left\{\hat{n}(\boldsymbol{r})\bar{\varrho}_s(t,0)\right\}, \tag{4.33a}$$

$$\boldsymbol{I}_n(\boldsymbol{r},t) = \mathrm{Tr}\left\{\hat{\boldsymbol{I}}_n(\boldsymbol{r})\bar{\varrho}_s(t,0)\right\}, \tag{4.33b}$$

$$I_n^{[r]}(\boldsymbol{r},t) = \mathrm{Tr}\left\{\hat{I}_n^{[r]}(\boldsymbol{r})\bar{\varrho}_s(t,0)\right\}, \tag{4.33c}$$

$$h(\boldsymbol{r},t) = \mathrm{Tr}\left\{\hat{h}(\boldsymbol{r})\bar{\varrho}_s(t,0)\right\}, \tag{4.33d}$$

$$\hat{\boldsymbol{I}}_h(\boldsymbol{r},t) = \mathrm{Tr}\left\{\hat{\boldsymbol{I}}_h(\boldsymbol{r})\bar{\varrho}_s(t,0)\right\}, \tag{4.33e}$$

$$I_h^{[r]}(\boldsymbol{r},t) = \mathrm{Tr}\left\{\hat{I}_h^{[r]}(\boldsymbol{r})\bar{\varrho}_s(t,0)\right\}, \tag{4.33f}$$

with $r = 2,3,\ldots$, and the right hand side of each equation (4.33) is a highly-nonlinear functional of the Lagrange multipliers of Eq. (4.28). For illustration let us consider the relation of fluxes and Lagrange multipliers in first order, that is, in a *linear approximation*. For that purpose we resort to Heims-Jaynes pertubation expansion for averages [179]. Firstly we write the NESO of Eq. (4.27) in the form

$$\bar{\varrho}_s(t,0) = \frac{\exp[\hat{A}+\hat{B}]}{\mathrm{Tr}\{\exp[\hat{A}+\hat{B}]\}}\,, \tag{4.34}$$

where

$$\begin{aligned}\hat{A} = &- F_h(t)\hat{H}_0 - F_n(t)\hat{N} - \boldsymbol{\alpha}_h \cdot \hat{\boldsymbol{I}}_h \\ &- \boldsymbol{\alpha}_n \cdot \hat{\boldsymbol{I}}_n - \sum_{r\geq 2}[\alpha_h^{[r]} \otimes \hat{I}_h^{[r]} + \alpha_n^{[r]} \otimes \hat{I}_n^{[r]}]\,,\end{aligned} \tag{4.35}$$

that is the contribution from the homogeneous part, that is, $Q = 0$ in Eq. (4.27), and $\hat{B}$ consists of the remaining terms, that is, those involving the inhomogeneous ($Q \neq 0$) contributions. Moreover, we define

$$\bar{\varrho}_0(t,0) = \frac{\exp[\hat{A}]}{\mathrm{Tr}\{\exp[\hat{A}]\}}\,, \tag{4.36}$$

that is, a distribution for the homogeous ($Q = 0$) state. Using the first order contribution in Heims-Jaynes expansion [179] it is found for the fluxes that

$$\begin{aligned}\boldsymbol{I}_h(\boldsymbol{Q},t) = &\Lambda_{hh}^{[2]}(\boldsymbol{Q},t) \otimes \boldsymbol{\alpha}_h(\boldsymbol{Q},t) + \Lambda_{hn}^{[2]}(\boldsymbol{Q},t) \otimes \boldsymbol{\alpha}_n(\boldsymbol{Q},t) \\ &+ \sum_{r\geq 2}\left[\Lambda_{hh}^{[r+1]}(\boldsymbol{Q},t) \otimes \alpha_h^{[r]}(\boldsymbol{Q},t) + \Lambda_{hn}^{[r+1]}(\boldsymbol{Q},t) \otimes \alpha_n^{[r]}(\boldsymbol{Q},t)\right]\,,\end{aligned} \tag{4.37a}$$

$$\begin{aligned}\boldsymbol{I}_n(\boldsymbol{Q},t) = &\Lambda_{nh}^{[2]}(\boldsymbol{Q},t) \otimes \boldsymbol{\alpha}_h(\boldsymbol{Q},t) + \Lambda_{nn}^{[2]}(\boldsymbol{Q},t) \otimes \boldsymbol{\alpha}_n(\boldsymbol{Q},t) \\ &+ \sum_{r\geq 2}\left[\Lambda_{nh}^{[r+1]}(\boldsymbol{Q},t) \otimes \alpha_h^{[r]}(\boldsymbol{Q},t) + \Lambda_{nn}^{[r+1]}(\boldsymbol{Q},t) \otimes \alpha_n^{[r]}(\boldsymbol{Q},t)\right]\,,\end{aligned} \tag{4.37b}$$

$$\begin{aligned}I_h^{[r]}(\boldsymbol{Q},t) = &\Lambda_{hh}^{[r]}(\boldsymbol{Q},t)F_h(\boldsymbol{Q},t) + \Lambda_{hn}^{[r]}(\boldsymbol{Q},t)F_n(\boldsymbol{Q},t) \\ &+ \Lambda_{hh}^{[r+1]}(\boldsymbol{Q},t) \otimes \boldsymbol{\alpha}_h(\boldsymbol{Q},t) + \Lambda_{hn}^{[r+1]}(\boldsymbol{Q},t) \otimes \boldsymbol{\alpha}_n(\boldsymbol{Q},t) \\ &+ \sum_{r'\geq 2}\left[\Lambda_{hh}^{[r+r']}(\boldsymbol{Q},t) \otimes \alpha_h^{[r']}(\boldsymbol{Q},t) + \Lambda_{hn}^{[r+r']}(\boldsymbol{Q},t) \otimes \alpha_n^{[r']}(\boldsymbol{Q},t)\right]\,,\end{aligned} \tag{4.37c}$$

$$
\begin{aligned}
I_n^{[r]}(\mathbf{Q},t) &= \Lambda_{nh}^{[r]}(\boldsymbol{Q},t)F_h(\boldsymbol{Q},t) + \Lambda_{nn}^{[r]}(\boldsymbol{Q},t)F_n(\boldsymbol{Q},t) \\
&\quad + \Lambda_{nh}^{[r+1]}(\boldsymbol{Q},t) \otimes \boldsymbol{\alpha}_h(\boldsymbol{Q},t) + \Lambda_{nn}^{[r+1]}(\boldsymbol{Q},t) \otimes \boldsymbol{\alpha}_n(\boldsymbol{Q},t) \\
&\quad + \sum_{r'\geq 2} \left[\Lambda_{nh}^{[r+r']}(\boldsymbol{Q},t) \otimes \alpha_h^{[r']}(\boldsymbol{Q},t) + \Lambda_{nn}^{[r+r']}(\boldsymbol{Q},t) \otimes \alpha_n^{[r']}(\boldsymbol{Q},t)\right] ,
\end{aligned}
\tag{4.37d}
$$

we recall that in these equations $\boldsymbol{Q} \neq \mathbf{0}$, and the Λ's are correlation-like functions derived as follows.

Taking into account the separation of the statistical operator as given in Eqs. (4.34) to (4.36), we obtain resorting, as said, to Heims-Jaynes expansion [179], when used up to first order only (meaning weak fluxes), that for example the first order (vectorial) flux of the energy density is

$$
\boldsymbol{I}_h(\boldsymbol{Q},t) = \langle \hat{\boldsymbol{I}}_h(\boldsymbol{Q})|t\rangle_0[1 - \langle \hat{B}|t\rangle_0] + \langle \hat{S}_1 \hat{\boldsymbol{I}}_h(\boldsymbol{Q})|t\rangle_0 , \tag{4.38}
$$

where subindex nought indicates average value using the partial statistical operator of Eq. (4.36), and where

$$
\hat{S}_1 = \int_0^1 dx\, \mathrm{e}^{-x\hat{A}}\hat{B}\, \mathrm{e}^{x\hat{A}} . \tag{4.39}
$$

But, the average value of the energy flux operator over the homogeneous state defined by $\bar{\varrho}_0$ is nil, and then

$$
\hat{\boldsymbol{I}}_h(\boldsymbol{Q},t) = \mathrm{Tr}\left\{\int_0^1 dx\, \mathrm{e}^{-x\hat{A}}\hat{B}\, \mathrm{e}^{x\hat{A}}\hat{\boldsymbol{I}}_h(\boldsymbol{Q})\bar{\varrho}_0(t,0)\right\} , \tag{4.40}
$$

which is of the form of Eq. (4.37a) where

$$
\Lambda_{hh}^{[2]}(\boldsymbol{Q},t) = -\,\mathrm{Tr}\left\{\sum_{\boldsymbol{Q}'} \int_0^1 dx \left[\hat{\boldsymbol{I}}_h(\boldsymbol{Q}',x)\hat{\boldsymbol{I}}_h(\boldsymbol{Q})\right] \bar{\varrho}_0(t,0)\right\} , \tag{4.41a}
$$

$$
\Lambda_{hn}^{[2]}(\boldsymbol{Q},t) = -\,\mathrm{Tr}\left\{\sum_{\boldsymbol{Q}'} \int_0^1 dx \left[\hat{\boldsymbol{I}}_n(\boldsymbol{Q}',x)\hat{\boldsymbol{I}}_h(\boldsymbol{Q})\right] \bar{\varrho}_0(t,0)\right\} , \tag{4.41b}
$$

$$
\Lambda_{hh}^{[r+1]}(\boldsymbol{Q},t) = -\,\mathrm{Tr}\left\{\sum_{\boldsymbol{Q}'} \int_0^1 dx \left[\hat{I}_h^{[r]}(\boldsymbol{Q}',x)\hat{\boldsymbol{I}}_h(\boldsymbol{Q})\right] \bar{\varrho}_0(t,0)\right\} , \tag{4.41c}
$$

$$
\Lambda_{hn}^{[r+1]}(\boldsymbol{Q},t) = -\,\mathrm{Tr}\left\{\sum_{\boldsymbol{Q}'} \int_0^1 dx \left[\hat{I}_n^{[r]}(\boldsymbol{Q}',x)\hat{\boldsymbol{I}}_h(\boldsymbol{Q})\right] \bar{\varrho}_0(t,0)\right\} . \tag{4.41d}
$$

Similarly, it follows for general $p, p' \in \{n, h\}$ and $r \geq 1, r' \geq 0$ that

$$\Lambda_{pp'}^{[r'+r]}(\mathbf{Q}, t) = -\operatorname{Tr}\Big\{\sum_{\mathbf{Q}'} \int_0^1 dx \left[\hat{I}_{p'}^{[r']}(\mathbf{Q}', x)\hat{I}_p^{[r]}(\mathbf{Q})\right] \bar{\varrho}_0(t, 0)\Big\}, \quad (4.42)$$

where $[\hat{I}_{p'}^{[r']}\hat{I}_p^{[r]}]$ stands for the product of tensors $\hat{I}_{p'}^{[r']}$ and $\hat{I}_p^{[r]}$ [177], and we have introduced the definition

$$\hat{I}_{p'}^{[r']}(\mathbf{Q}', x) = e^{-x\hat{A}}\hat{I}_{p'}^{[r']}(\mathbf{Q}')e^{x\hat{A}} . \quad (4.43)$$

It can be noticed that because of symmetry (Curie-Prigogine theorem) the kinetic coefficients of Eq. (4.42) are null except if both tensor ranks r and r' have the same parity (both even or both odd).

Equations (4.37) are very cumbersome expressions, but they are linear relations between fluxes and the corresponding set of Lagrange parameters. Hence, these relations can be inverted — a formidable task — and consequently the Lagrange multipliers, $\boldsymbol{\alpha}_h$, $\boldsymbol{\alpha}_n$, $\{\alpha_h^{[r]}\}$, $\{\alpha_n^{[r]}\}$, are linear combinations of the set of fluxes $\{\{\boldsymbol{I}_h, \boldsymbol{I}_n, \{I_h^{[r]}\}, \{I_n^{[r]}\}\}$, and of F_h and F_n. Since the latter satisfy the kinetic equations, Eqs. (4.31), we can derive equations of evolution for the associated Lagrange multipliers. Therefore, they are completely determined and their evolution in time given; and, consequently, they completely define the macroscopic state of the system. Furthermore, they can be measured via an experiment, which can be described within the NESOM response function theory [101, 175, 180]. Several examples are given in references [151, 152, 155, 158, 176]. In continuation we present brief considerations on the behavior of these fluxes.

4.4 The nonequilibrium thermodynamics with fluxes (grand-canonical)

As discussed in the previous Chapter, IST has an accompanying state-like function consisting in the so-called informational entropy, given by

$$\bar{S}(t) = -\operatorname{Tr}\Big\{\varrho_\epsilon(t)\mathcal{P}_\varepsilon(t)\ln\varrho_\epsilon(t)\Big\}, \quad (4.44)$$

where $\mathcal{P}_\varepsilon(t)$ is the time-dependent projection operator, defined in Ref. [101], which projects over the subspace generated by the basic dynamical variables (referred to as the informational subspace, sometimes also referred to as the space of relevant variables [181]) and depending at each time on the macroscopic nonequilibrium state of the system, i.e. $\mathcal{P}_\varepsilon(t)\ln\varrho_\epsilon(t) = \ln\bar{\varrho}(t,0)$ [64,65,101,137] [see Eq. (2.18)]. A geometrical-topological description is given by Balian et al. [181]. As already noticed in the preceding Chapter, this entropy-like function defined in the context of IST satisfies a $\mathcal{H}$-like theorem; generalizations of Glansdorff-Prigogine criteria of evolution and (in)stability [166]; whose space and time dependent density, $\eta(\mathbf{r},t)$ of Eq. (3.2) satisfies the continuity equation Eq. (3.3), that is

$$\boxed{\frac{\partial}{\partial t}\bar{\eta}(\mathbf{r},t) + \nabla\cdot\mathbf{I}_\eta(\mathbf{r},t) = \sigma_\eta(\mathbf{r},t)\,,} \tag{4.45}$$

where, we recall, $\mathbf{I}_\eta$ is the informational entropy flux and σ_η the density of the informational entropy production: In the description given by the set of Eq. (4.30), and in the direct space, we have that in Eq. (4.45)

$$\begin{aligned}\mathbf{I}_\eta(\mathbf{r},t) &= F_h(\mathbf{r},t)\mathbf{I}_h(\mathbf{r},t) + F_n(\mathbf{r},t)\mathbf{I}_n(\mathbf{r},t)\\ &\quad+\sum_{r\geq 2}\left[F_h^{[r-1]}(\mathbf{r},t)\otimes I_h^{[r]}(\mathbf{r},t) + F_n^{[r-1]}(\mathbf{r},t)\otimes I_n^{[r]}(\mathbf{r},t)\right],\end{aligned} \tag{4.46}$$

which, it can be noticed, is a sum of the products of the fluxes with the Lagrange multiplier of the preceding flux (those for the densities in the case of the vectorial fluxes), and

$$\begin{aligned}\sigma_\eta(\mathbf{r},t) &= \nabla F_h(\mathbf{r},t)\cdot\mathbf{I}_h(\mathbf{r},t) + \nabla F_n(\mathbf{r},t)\cdot\mathbf{I}_n(\mathbf{r},t)\\ &\quad+\sum_{r\geq 2}\left[\nabla F_h^{[r-1]}(\mathbf{r},t)\otimes I_h^{[r]}(\mathbf{r},t) + \nabla F_n^{[r-1]}(\mathbf{r},t)\otimes I_n^{[r]}(\mathbf{r},t)\right]\\ &\quad+F_h(\mathbf{r},t)\mathcal{J}_h(\mathbf{r},t) + F_n(\mathbf{r},t)\mathcal{J}_n(\mathbf{r},t)\\ &\quad+\boldsymbol{\alpha}_h(\mathbf{r},t)\cdot\mathbf{J}_h(\mathbf{r},t) + \boldsymbol{\alpha}_n(\mathbf{r},t)\cdot\mathbf{J}_n(\mathbf{r},t)\\ &\quad+\sum_{r\geq 2}\left[\alpha_h^{[r]}(\mathbf{r},t)\otimes\mathcal{J}_h^{[r]}(\mathbf{r},t) + \alpha_n^{[r]}(\mathbf{r},t)\otimes\mathcal{J}_n^{[r]}(\mathbf{r},t)\right].\end{aligned} \tag{4.47}$$

Moreover, there follows the relevant result that

$$F_h(\mathbf{r},t) = \frac{\delta\bar{S}(t)}{\delta h(\mathbf{r},t)}\,;\qquad F_n(\mathbf{r},t) = \frac{\delta\bar{S}(t)}{\delta n(\mathbf{r},t)}\,; \tag{4.48a}$$

$$\boldsymbol{\alpha}_h(\mathbf{r},t) = \frac{\delta\bar{S}(t)}{\delta\mathbf{I}(\mathbf{r},t)}\,;\qquad \boldsymbol{\alpha}_n(\mathbf{r},t) = \frac{\delta\bar{S}(t)}{\delta\mathbf{I}_n(\mathbf{r},t)}\,; \tag{4.48b}$$

$$\alpha_h^{[r]}(\mathbf{r},t) = \frac{\delta \bar{S}(t)}{\delta I_h^{[r]}(\mathbf{r},t)} \; ; \qquad \alpha_n^{[r]}(\mathbf{r},t) = \frac{\delta \bar{S}(t)}{\delta I_n^{[r]}(\mathbf{r},t)} \, , \tag{4.48c}$$

or, in words, the Lagrange multipliers (intensive nonequilibrium thermodynamics variables) are the differential coefficients (here in terms of a functional differential [178]) of the informational entropy with respect to the associated specific variables: These are the *nonequilibrium thermodynamic equations of state* in IST [also cf. Eqs. (4.29)].

Keeping an analogy with equilibrium and local equilibrium theories, we can introduce alternative forms for the Lagrange multipliers, namely

$$F_h(\mathbf{r},t) = \frac{1}{k_B \Theta(\mathbf{r},t)} \, , \tag{4.49a}$$

defining the space and time dependent variable Θ, designated as quasitemperature [166, 176];

$$F_n(\mathbf{r},t) = -\frac{\mu(\mathbf{r},t)}{k_B \Theta(\mathbf{r},t)} \, , \tag{4.49b}$$

introducing the space and time dependent variable μ, designated as a local quasi-chemical potential [176];

$$\boldsymbol{\alpha}_h(\mathbf{r},t) = -\frac{\mathbf{v}_h(\mathbf{r},t)}{k_B \Theta(\mathbf{r},t)} \, , \tag{4.49c}$$

introducing a drift-velocity field $\mathbf{v}_h$ for the energy density;

$$\boldsymbol{\alpha}_n(\mathbf{r},t) = -\frac{\mathbf{v}_n(\mathbf{r},t)}{k_B \Theta(\mathbf{r},t)} \, , \tag{4.49d}$$

with $\mathbf{v}_n$ being a drift-velocity field for the particles; and

$$\alpha_h^{[r]}(\mathbf{r},t) = -\frac{v_h^{[r]}(\mathbf{r},t)}{k_B \Theta(\mathbf{r},t)} \, , \tag{4.49e}$$

$$\alpha_n^{[r]}(\mathbf{r},t) = -\frac{v_n^{[r]}(\mathbf{r},t)}{k_B \Theta(\mathbf{r},t)} \, , \tag{4.49f}$$

introducing the quantities $v_{h(n)}^{[r]}$, which are rank $r \geq 2$ tensors, associated to the higher order fluxes of energy and particle densities which are also tensors of rank r.

Resorting to the definitions given in these Eqs. (4.49), we can rewrite the IST entropy flux of Eq. (4.46) as

$$I_\eta(\boldsymbol{r},t) = -\frac{\boldsymbol{I}_q(\boldsymbol{r},t)}{k_B\Theta(\boldsymbol{r},t)}\,, \tag{4.50}$$

which defines the heat flux in IST, given by

$$\begin{aligned}\boldsymbol{I}_q(\boldsymbol{r},t) = \boldsymbol{I}_h(\boldsymbol{r},t) - \mu(\boldsymbol{r},t)\boldsymbol{I}_n(\boldsymbol{r},t)\\ - \boldsymbol{v}_h(\boldsymbol{r},t)\otimes I_h^{[2]}(\boldsymbol{r},t) - \boldsymbol{v}_n(\boldsymbol{r},t)\otimes I_n^{[2]}(\boldsymbol{r},t)\\ - \sum_{r\geq 3}\left[v_h^{[r-1]}(\boldsymbol{r},t)\otimes I_h^{[r]}(\boldsymbol{r},t) + v_n^{[r-1]}(\boldsymbol{r},t)\otimes I_n^{[r]}(\boldsymbol{r},t)\right]\,.\end{aligned} \tag{4.51}$$

All the expressions given above, in the context of IST, for entropy density, entropy-production density, entropy flux, equations of state, and heat flux, go over the corresponding well known values in classical (Onsagerian) Irreversible Thermodynamics when the proper asymptotic limit is taken, namely all Lagrange multipliers α are taken as null, and the equations of evolution for the fluxes are ignored and replaced by Fourier and Fick constitutive equations. Moreover, also introducing additional restrictive conditions the results of earlier versions of Extended Irreversible Thermodynamics [25] are recovered. In the next section we present an example for illustration.

4.5 Thermodynamics of a fermion system

Let us consider the electron subsystem in a n-doped semiconductor [151]. These electrons are treated as Landau quasi-particles in the random phase approximation, and the effective mass approximation is used (meaning that in Eq. (4.5) $\varepsilon_{k\sigma} = \hbar^2k^2/2m^*$) [182]. Moreover, they are assumed to be in statistical-nondegenerate-like states, a good approximation in most experimental conditions.

We consider next the thermo-hydrodynamics of this fermion system in the framework of the theory so far presented, but introducing a truncation (evaluation of the truncation procedure is discussed in [133–135]) in the choice of the basic set of variables, i.e. Zubarev-Peletminskii closure condition [of Eq. (2.4)] is violated. The one we introduce consists into keeping only the densities of energy and particles, $h(\boldsymbol{r},t)$ and $n(\boldsymbol{r},t)$, and their first fluxes, $\boldsymbol{I}_h(\boldsymbol{r},t)$, $\boldsymbol{I}_n(\boldsymbol{r},t)$, and, of course, their associated Lagrange multipliers, $F_h(\boldsymbol{r},t)$, $F_n(\boldsymbol{r},t)$, $\boldsymbol{\alpha}_h(\boldsymbol{r},t)$, $\boldsymbol{\alpha}_n(\boldsymbol{r},t)$, respectively (or $\Theta(\boldsymbol{r},t)$, $\mu(\boldsymbol{r},t)$, $\boldsymbol{v}_h(\boldsymbol{r},t)$, $\boldsymbol{v}_n(\boldsymbol{r},t)$ if the definitions introduced in Eqs. (4.49) are used). The auxiliary nonequilibrium statistical operator is the one of Eq. (4.27) but setting all $\alpha_h^{[r]}$ and $\alpha_n^{[r]}$ for $r \geq 2$ equal to zero. Moreover, the electrons interact with the lattice vibrations, and since we are considering polar semiconductors (GaAs will be chosen for numerical calculations) we retain only the predominant Fröhlich interaction [182, 183] between electrons and longitudinal optical phonons. Furthermore, we assume the LO phonons to be constantly kept in equilibrium with the acoustic phonons, with the latter at temperature T_0. We notice that, because of the use of the effective mass approximation, the quantity of Eq. (4.15), which is fundamental in the definition of the fluxes, takes in this case the simple expression

$$\boldsymbol{u}(\boldsymbol{k},\boldsymbol{Q},\sigma) = \frac{\hbar \boldsymbol{k}}{m^*} , \tag{4.52}$$

and then

$$\boldsymbol{I}_h(\boldsymbol{Q}) = \sum_{\boldsymbol{k}\sigma} \varepsilon_{\boldsymbol{k}\sigma} \frac{\hbar \boldsymbol{k}}{m^*} \hat{n}_{\boldsymbol{k}\boldsymbol{Q}\sigma} ; \tag{4.53}$$

$$\boldsymbol{I}_n(\boldsymbol{Q}) = \sum_{\boldsymbol{k}\sigma} \frac{\hbar \boldsymbol{k}}{m^*} \hat{n}_{\boldsymbol{k}\boldsymbol{Q}\sigma} . \tag{4.54}$$

The analysis is further restricted to introducing a decoupling of particle and energy motion (this is equivalent, in this case of charged particles, to disregard cross-kinetic terms consisting in thermo-electric effects, which are a result that the motion of the charged particles imply in an electric current with an accompanying energy current). We consider next the *presence of weak inhomogeneities travelling on a background consisting of a steady and homogeneous state.* As a consequence we shall use a linear approximation in the amplitudes of the inhomogeneities, and we write for the quantities of Eqs. (4.49a) and (4.49b)

$$\Theta(\boldsymbol{r},t) = T^* + \Delta\Theta(\boldsymbol{r},t) , \tag{4.55a}$$

$$\mu(\boldsymbol{r},t) = \mu^* + \Delta\mu(\boldsymbol{r},t) , \tag{4.55b}$$

where T^* is the constant quasitemperature and μ^* is the constant quasi-chemical potential of the steady and homogenous state; $\Delta\Theta$ ($\ll T^*$) and $\Delta\mu$ ($\ll \mu^*$) are the space and time dependent departures from it. In this case the statistical operator can be split into two contributions as in Eqs. (4.34) to (4.36), but now involving the one relative to the homogeneous state, which, we recall, it is also considered as stationary (for example a plain equilibrium state, or a nonequilibrium one arising out of the action of a constant external excitation-pumping mechanism), given by

$$\bar{\varrho}_0 = \exp\left\{-\phi_0 - F_h\hat{H}_0 - F_n\hat{N}\right\} \equiv \frac{\exp\left[\hat{A}\right]}{\mathrm{Tr}\left\{\exp\left[\hat{A}\right]\right\}}, \tag{4.56}$$

and the part associated to the inhomogeneous contribution is

$$\begin{aligned}\hat{B} = -\sum_{Q\neq 0}\Big[&F_h(Q,t)\hat{h}(Q) + F_n(Q,t)\hat{n}(Q) + \\ &+ \boldsymbol{\alpha}_h(Q,t)\cdot\hat{\boldsymbol{I}}_h(Q) + \boldsymbol{\alpha}_h(Q,t)\cdot\hat{\boldsymbol{I}}_n(Q)\Big],\end{aligned} \tag{4.57}$$

and we have assumed null global fluxes (i.e. absence of homogeneous currents of particle and energy). The equations of evolution for the energy density and its flux are given by Eqs. (4.31a) and (4.31c) and for the particle-number density and its flux by Eqs. (4.31b) and (4.31d). But in the truncated representation we are using, we need to express the second order flux of energy density (particle-number density) and the collision operators in terms of h (n) and $\boldsymbol{I}_h$ ($\boldsymbol{I}_n$) only. The collision operators are dealt with in the Markovian approximation [184] (second order in Fröhlich interaction strength), and after some algebra it is found, in the given conditions, considering only the motion of the energy density as decoupled of the motion of the particle-number density, that

$$\frac{\partial}{\partial t}h(\boldsymbol{r},t) + \nabla\cdot\boldsymbol{I}_h(\boldsymbol{r},t) = -\theta_h^{-1}h(\boldsymbol{r},t), \tag{4.58a}$$

$$\frac{\partial}{\partial t}\boldsymbol{I}_h(\boldsymbol{r},t) + \frac{1}{3}v_{th}^2\nabla h(\boldsymbol{r},t) = -\theta_I^{-1}\boldsymbol{I}_h(\boldsymbol{r},t), \tag{4.58b}$$

where the coefficients θ_h and θ_I play the role of relaxation times and $m^* v_{th}^2/2 = 3k_B T^*/2$ (v_{th} being the so-called thermal velocity). In Eq. (4.58b), θ_I can be interpreted as Maxwell's relaxation time [35]; we omit to write down their detailed expressions, which are given in [160].

In Eqs. (4.58a) and (4.58b) both relaxation times are constant, because, in the linear approximation we are using, they are given as correlation functions defined over the homogeneous and stationary reference state. Using Heims-Jaynes expansion in its lowest (linear) order [179] we can relate the density of energy and the Lagrange multiplier F_h (or the quasitemperature Θ) in the form

$$h(\boldsymbol{r},t) = \mathrm{Tr}\left\{\hat{h}(\boldsymbol{r})\bar{\varrho}(t,0)\right\} \simeq -l_h F_h(\boldsymbol{r},t) , \tag{4.59}$$

where $l_h = (15/8)\bar{n}(k_B T^*)^2$, with $\bar{n}$ being the particle density, and then [cf. Eq. (4.49a)]

$$\nabla h(\boldsymbol{r},t) \simeq \frac{k_B l_h}{(k_B T^*)^2} \nabla\Theta(\boldsymbol{r},t) , \tag{4.60}$$

and

$$-\theta_I \frac{\partial}{\partial t} \boldsymbol{I}_h(\boldsymbol{r},t) - \kappa\nabla\Theta(\boldsymbol{r},t) = \boldsymbol{I}_h(\boldsymbol{r},t) , \tag{4.61}$$

where $\kappa = k_B v_{th}^2 l_h \theta_I / 3(k_B T^*)^2 = 15\bar{n}k_B^2 T^* \theta_I / 4m^*$. Therefore, if in this strongly truncated version we have introduced we further assume a weak change in time of the energy flux, i.e. $\partial \boldsymbol{I}_h/\partial t \simeq 0$, implying in a near stationary flux motion, there follows a local-in-space Fourier law, namely

$$\boldsymbol{I}_h(\boldsymbol{r},t) = -\kappa\nabla\Theta(\boldsymbol{r},t) , \tag{4.62}$$

that is, the constitutive equation of classical (linear) irreversible thermodynamics, with κ playing the role of the heat conductivity. It can be noticed that the energy flux above coincide with the heat flux because the material motion has been disregarded, otherwise the heat flux is given by $\boldsymbol{I}_q(\boldsymbol{r},t) = \boldsymbol{I}_h(\boldsymbol{r},t) - \mu(\boldsymbol{r},t)\boldsymbol{I}_n(\boldsymbol{r},t)$.

Returning to Eqs. (4.58), differentiating in time Eq. (4.58a), and using Eqs. (4.58b), (4.59), and (4.49a), it follows an equation of evolution for the quasitemperature in the form

$$\left[\frac{1}{c_h^2}\frac{\partial^2}{\partial t^2} + \frac{1}{D_h}\frac{\partial}{\partial t} - \nabla^2\right]\Theta(\boldsymbol{r},t) = -\frac{\Theta(\boldsymbol{r},t)}{\Lambda^2} , \tag{4.63}$$

which is of the type of the hyperbolic-type telegraphist equation of classical electrodynamics, but with an additional term on the right. In this

Eq. (4.63), $c_h = v_{th}/\sqrt{3}$ is the velocity of propagation in the motion of energy, $\Lambda^2 = c_h^2 \theta_h \theta_I$, and $D_h = c_h^2 \theta_* = v_{th}^2 \theta_*/3$ (with $\theta_*^{-1} = \theta_h^{-1} + \theta_I^{-1}$) is a diffusion coefficient. This equation, once we disregard the right hand side, is also of the form of the one that is derived within the framework of phenomenological Extended Irreversible Thermodynamics [25].

Equation (4.63) implies in *second sound* propagation in the carrier system. As known [185], such motion is a superposition of diffusive motion (composed of the long wavelengths) and damped undulatory motion (at intermediate to short wavelengths); these characteristics are evidenced in experiments related to the techno-industrial process of thermal stereolithography [186]. Furthermore, if in Eq. (4.63) we take c_h going to infinity while D_h is kept constant, and the right side is disregarded, we obtain that

$$\frac{\partial}{\partial t}\Theta(\boldsymbol{r},t) - D_h \nabla^2 \Theta(\boldsymbol{r},t) = 0\,, \tag{4.64}$$

namely Fourier's heat diffusion equation, which is derived in classical (Onsagerian) Irreversible Thermodynamics from the constitutive equation, Eq. (4.62).

The results described above indicate that Classical Irreversible Thermodynamics and earlier versions of Extended Irreversible Thermodynamics are retrieved, in this statistical approach, after using quite stringent asymptotic limits of IST: The first requires to introduce the densities of particle-number and energy plus postulated constitutive equations; the second introduces these densities and their fluxes (plus locality in space and memoryless processes), and, therefore, both imply in radical truncations in the closure condition required by IST, accompanied with processes of linearization, Markovianization, and of neglecting spatial correlations. On this question, we call the attention to the Figs. 1 and 2 in [109], which we reproduce here as Figs. 4.1 and 4.2, where it is shown that when smoother the variation in space of the densities (that is, predominance of the longest wavelengths) the smaller the dimension of the state space to be chosen (i.e. the more stringent the truncation). In Fig. 4.1 is described the transition from the classical to the first extended description (the frontier wavenumber is indicated in the figure), and an experimental corroboration may follow from the analysis of the Raman spectra as shown in Fig. 4.2: When going from the case of propagation of second sound to the diffusive regime, the Brillouin doublet, associated to second sound propagation, should collapse into a shiftless Rayleigh band

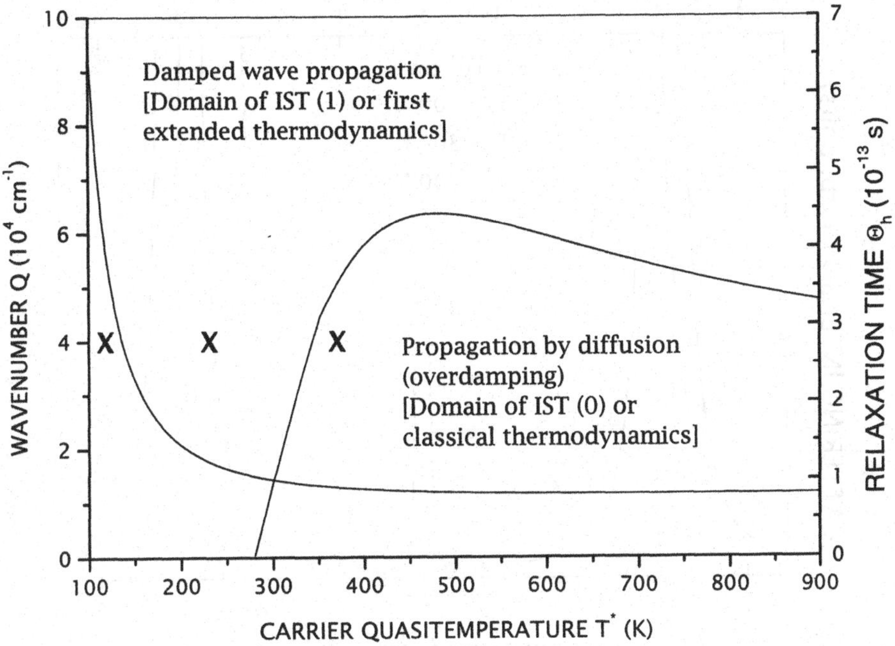

Figure 4.1: The full line separates the regions of values of the wavenumber (left ordinate), which, for each quasitemperature, correspond to either damped second sound propagation (upper part) or to diffusive (Fourier-type) movement (lower and right part). The dashed line provides the values of the relaxation time defined in the main text. After reference [109].

corresponding to purely diffusive motion. We note that in Fig. 4.1 we have characterized the state of the carrier system by the quasitemperature T^*, meaning, we recall, that we have written $\Theta(\boldsymbol{r}, t) = T^* + \Delta\Theta(\boldsymbol{r}, t)$, where T^* is the quasitemperature of the homogeneous and stationaty state of reference and $\Delta\Theta \ll T^*$.

4.6 Some Properties of the Informational-Statistical Entropy

Let us consider the continuity equation for the MaxEnt-NESOM entropy of Eq. (4.45), where, we recall, $\boldsymbol{I}_\eta$ is the informational-entropy flux, given by

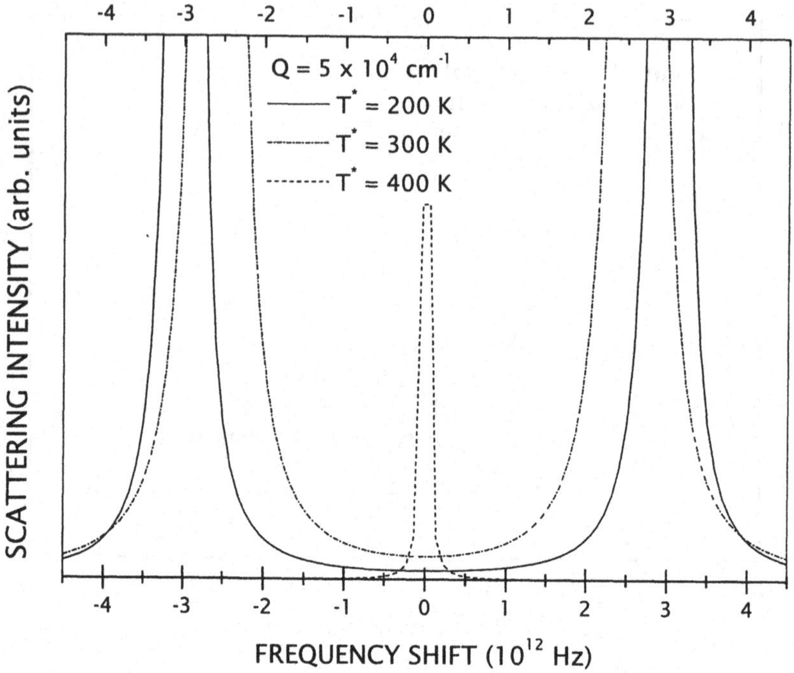

Figure 4.2: Expected Raman scattering spectra by carriers corresponding to the positioning of the three crosses in Fig. 4.1. After reference [109].

Eq. (4.46), and σ_η the density of the informational-entropy production, given by Eq. (4.47).

Integrating in space Eq. (4.45), we obtain an expression for the global informational-entropy production, namely

$$\bar{\sigma}(t) = \frac{d\bar{S}(t)}{dt} = \int d^3r \frac{\partial \bar{\eta}(\boldsymbol{r},t)}{\partial t} = \int d^3r \sigma_\eta(\boldsymbol{r},t) - \int d^3r \nabla \cdot \boldsymbol{I}_\eta(\boldsymbol{r},t) \,. \tag{4.65}$$

But, using Gauss theorem in the last integral of Eq. (4.65), we can rewrite it as

$$\frac{d\bar{S}(t)}{dt} = \int d^3r \sigma_\eta(\boldsymbol{r},t) - \int_\Sigma d\boldsymbol{\Sigma} \cdot \boldsymbol{I}_\eta(\boldsymbol{r},t) \,, \tag{4.66}$$

where the last term is a surface integral over the system boundaries. Equation (4.66) allows us to separate the entropy production into two contributions (as is usually done, cf. [20] and [169]): one is the *internal production of informational entropy*

$$P(t) = \int d^3r\,\sigma_\eta(\mathbf{r},t)\,, \tag{4.67}$$

and the other is the *exchange of informational entropy with the surroundings,* or *flux term*

$$P_\Sigma(t) = -\int_\Sigma d\Sigma \cdot \mathbf{I}_\eta(\mathbf{r},t) = -\sum_{j=1}^{n}\int_\Sigma d\Sigma \cdot \sum_{r\geq 1} F_j^{[r-1]}(\mathbf{r},t) \otimes I_j^{[r]}(\mathbf{r},t)\,. \tag{4.68}$$

Furthermore, using Eq. (4.47), Eq. (4.67) becomes

$$P(t) = \bar{\sigma}(t) + \int d^3r \sum_{j=1}^{n}\sum_{r\geq 1} \nabla F_j^{[r-1]}(\mathbf{r},t) \otimes I_j^{[r]}(\mathbf{r},t)\,, \tag{4.69}$$

and then taking into account that

$$\bar{\sigma}(t) = \frac{d\bar{S}(t)}{dt} = P(t) + P_\Sigma(t)\,, \tag{4.70}$$

use of Eqs. (4.68) and (4.69) implies in that

$$\sum_{j=1}^{n}\sum_{r\geq 1}\int d^3r \nabla F_j^{[r-1]}(\mathbf{r},t) \otimes I_j^{[r]}(\mathbf{r},t) = -P_\Sigma(t)\,. \tag{4.71}$$

This is just a consequence that the quantities $J_j^{(0)}$ (which gave origin to the divergence of the flux) and $J_j^{(1)}$ do not contribute to the informational-entropy production, what is shown in Appendix II, that is, as already noticed, there is no dissipation in the nonequilibrium macroscopic description associated to the auxiliary distribution $\bar{\varrho}(t,0)$, but the irreversible processes are fully accounted for by $\varrho'_\varepsilon(t)$ [cf. Eq. (2.17)]. Explicitly, once the equations of evolution, Eqs. (4.31), are taken into account,

$$\bar{\sigma}(t) = \int d^3r \frac{\partial}{\partial t}\eta(\mathbf{r},t) = -\int d^3r \nabla \cdot \mathbf{I}_\eta(\mathbf{r},t) + \sum_{r\geq 0}\sum_{p}\int d^3r F_p^{[r]}(\mathbf{r},t) \otimes \mathcal{J}_p^{[r]}(\mathbf{r},t) \tag{4.72}$$

with $p = n$ and h. But with null flux of entropy over the system boundaries (see Appendix II)

$$\int d^3r\nabla \cdot \boldsymbol{I}_\eta(\boldsymbol{r},t) = -\int d\Sigma \cdot \boldsymbol{I}_\eta(\boldsymbol{r},t) = 0\,, \tag{4.73}$$

and then there survives the last term on the right of Eq. (4.72), that is

$$\bar{\sigma}(t) = \sum_{r\geq 0}\sum_{p}\int d^3r F_p^{[r]}(\boldsymbol{r},t)\otimes \mathrm{Tr}\Big\{\frac{1}{i\hbar}\left[\hat{I}_p^{[r]}(\boldsymbol{r}),\hat{H}'\right]\varrho'_\varepsilon(t)\Big\}\,. \tag{4.74}$$

Clearly $\bar{\sigma}(t)$ is zero for null ϱ'_ε, what is also the case when H' is null.

Summarizing these results, the informational-entropy production density $\sigma_\eta(\boldsymbol{r},t)$ of Eq. (4.47) accounts for the local internal production of informational entropy as indicated by Eq. (4.67). Moreover, using Eqs. (4.67) and (4.68), together with Eq. (4.70) we can make the identifications

$$\bar{\sigma}_{(i)}(t) = P(t)\,; \tag{4.75}$$

$$\bar{\sigma}_{(e)}(t) = P_\Sigma(t)\,. \tag{4.76}$$

Accordingly, Eq. (4.75) stands for the global internal production of informational-entropy while Eq. (4.76) is associated to informational-entropy exchange with the surroundings. In the appropriate limit (the truncation leading to the linear, or Onsagerian, approximation) the informational-entropy production $\bar{\sigma}(t)$ can then be interpreted as corresponding to the entropy production function of Classical Irreversible Thermodynamics.

4.7 The nonequilibrium equations of state

Let us next proceed to analize the differential coefficients of the informational entropy [cf. Eqs. (4.49)]. Consider a system composed of $\underline{s}$ subsystems. Let $\varepsilon_l(\boldsymbol{r},t)$ be the locally-defined energy densities and $n_l(\boldsymbol{r},t)$ the number densities in each $l(=1,2,\ldots,s)$ subsystem, which are taken as basic variables in MaxEnt-NESOM. We call $\beta_l(\boldsymbol{r},t)$ and $\zeta_l(\boldsymbol{r},t)$ their associated intensive variables [the Lagrange multipliers in Eq. (4.28) (but now for each kind of subsystem) that the formalism introduces]. But, as has already been shown (see also [134,187,188]), the selection rule (which encompasses the principle of equipresence [22]) of Eq. (2.4) requires the introduction of the fluxes of these quantities as basic variables, and with them all the other higher order fluxes (of tensorial rank $r \geq 2$). Consequently, the nonequilibrium statistical operator depends on all the densities and their fluxes, and Eq. (4.48) tell us that the Lagrange multipliers associated to them depend, each one, on all these basic variables, namely, the densities $\varepsilon_l(\boldsymbol{r},t)$, $n_l(\boldsymbol{r},t)$, their vectorial fluxes $\boldsymbol{I}_{\varepsilon l}(\boldsymbol{r},t)$, $\boldsymbol{I}_{nl}(\boldsymbol{r},t)$, and rank $r \geq 2$ tensorial fluxes $I_{\varepsilon l}^{[r]}(\boldsymbol{r},t)$, $I_{nl}^{[r]}(\boldsymbol{r},t)$.

The corresponding auxiliary coarse-grained ("instantaneously frozen") distribution is

$$\bar{\varrho}_s(t,0) = \exp\Big\{-\phi(t) - \sum_l \int d^3r[\beta_l(\boldsymbol{r},t)\hat{\varepsilon}_l(\boldsymbol{r}) + \\ + \zeta_l(\boldsymbol{r},t)\hat{n}_l(\boldsymbol{r}) + \boldsymbol{\alpha}_{\varepsilon l}(\boldsymbol{r},t)\cdot\hat{\boldsymbol{I}}_{\varepsilon l}(\boldsymbol{r}) + \boldsymbol{\alpha}_{nl}(\boldsymbol{r},t)\cdot\hat{\boldsymbol{I}}_{nl}(\boldsymbol{r}) \\ + \sum_{r\geq 2}\Big[\alpha_{\varepsilon l}^{[r]}(\boldsymbol{r},t)\otimes\hat{I}_{\varepsilon l}^{[r]}(\boldsymbol{r}) + \alpha_{nl}^{[r]}(\boldsymbol{r},t)\otimes\hat{I}_{nl}^{[r]}(\boldsymbol{r})\Big]\Big\}\,, \quad (4.77)$$

where the upper triangular hat stands for the dynamical operator of the corresponding quantity: the energy density $\hat{\varepsilon}$, the particle density $\hat{n}$, and their fluxes to all order $r \geq 1$ (r is also the corresponding tensorial rank, with $r = 1$ standing for the vectorial fluxes, which have been explicitly separated out in Eq. (4.77), and we have introduced β, ζ, and the α's as the corresponding Lagrange multipliers, which are the intensive nonequilibrium thermodynamic variables in IST for this case). We further recall that the proper nonequilibrium statistical operator that describes the dissipative irreversible macrostate of the system is $\varrho_\varepsilon(t)$ of Eq. (2.12), built on the basis of the auxiliary one of Eq. (4.77).

Thus, the Lagrange multipliers in MaxEnt-NESOM that are the differential coefficients of the informational entropy in IST (as already noted they are in this sense nonequilibrium thermodynamics variables conjugated to the basic ones), are for the densities

$$\beta_l(\boldsymbol{r},t) = \delta\bar{S}(t)/\delta\varepsilon_l(\boldsymbol{r},t)\,, \quad (4.78)$$

$$\zeta_l(\boldsymbol{r},t) = \delta\bar{S}(t)/\delta n_l(\boldsymbol{r},t)\,, \quad (4.79)$$

where δ, we recall, stands for functional differential [178]. These Eqs. (4.78) and (4.79), together with the equations for the functional derivatives of the informational statistical entropy with regard to the fluxes, namely $\boldsymbol{\alpha}_{\varepsilon l}(\boldsymbol{r},t) = \delta\bar{S}/\delta I_{\varepsilon l}(\boldsymbol{r},t)$, etc., constitute the set of nonequilibrium equations of state: They express each Lagrange multiplier (or differential coefficient of the informational-statistical entropy) to the set of basic macrovariables.

The IST informational entropy of Eqs. (3.3), appropriately given in terms of ε_l, n_l, and their fluxes of all order, goes over the corresponding one of local equilibrium in Classical Irreversible Thermodynamics, when all Lagrange multipliers β_l become identical for all subsystems and equal to the reciprocal of the local equilibrium temperature, while the ζ_l become equal to $-\beta\mu_l$, where μ_l are the local chemical potentials for the different chemical species in the material. All the other Lagrange multipliers, that is those associated to the fluxes, are null in such limit, and the equations of evolution are closed introducing *ad hoc* Fick and Fourier equations for the fluxes.

Consequently, in MaxEnt-NESOM we can introduce the space and time dependent nonequilibrium temperature-like variables $\Theta_l(\boldsymbol{r},t)$, which we call *quasitemperature for each subsystem* $l = 1$ *to* s, namely

$$\beta_l(\boldsymbol{r},t) = \delta\bar{S}(t)/\delta\varepsilon_l(\boldsymbol{r},t) \equiv \Theta_l^{-1}(\boldsymbol{r},t)\,, \tag{4.80}$$

where then (with $k = 1, 2, \ldots, s$)

$$\beta_l\{\varepsilon_k, n_k, \boldsymbol{I}_{\varepsilon k}, \boldsymbol{I}_{nk}, \{I_{\varepsilon k}^{[r]} r\}, \{I_{nk}^{[r]}\}\} \equiv \Theta_l^{-1}(\boldsymbol{r},t)\,, \tag{4.81}$$

with Boltzmann constant taken as unit. This Eq. (4.81) tells us that the Lagrange multiplier β_l represents in the MaxEnt approach to the NESOM a reciprocal quasitemperature (nonequilibrium temperature-like variable) of each subsystem, and which is dependent on the space and time coordinates. Evidently, as we have intended to indicate between the curly brackets in Eq. (4.81), it depends not only on the energies ε_l and densities n_l ($l = 1, 2, \ldots, s$) but on *all* the other basic variables which are appropriate for the description of the macroscopic state of the system, determined by the selection rule of Eq. (2.4), that is, the vectorial and tensorial fluxes. The definition of Eq. (4.80) properly recovers as particular limiting cases the local-equilibrium temperature of classical irreversible thermodynamics and the usual absolute temperature in equilibrium; in both cases there follows a unique temperature for all subsystems as it should.

A quite important aspect of the question needs be stressed at this point: Equation (4.80) is the formal definition of the so-called quasitemperature in IST, a very convenient one because of the analogy with local-equilibrium and equilibrium theories, which are recovered in the appropriate asymptotic limits. But we recall that it is a Lagrange multiplier that the method introduces from the outset, which is explicitly defined by the set of Eqs. (4.33), being a functional of the basic set of macrovariables. Therefore, its evolution in time, and then its local and instantaneous value, follows from the solution of the generalized transport equations, namely Eqs. (4.31), for the densities and all their fluxes. Thus, in IST, the quasitemperature of each subsystem is, as already emphasized, a *functional of all the basic variables,* which are, we insist, the densities and their fluxes of, in principle, all orders.

This question is extensively dealt with in references [174, 189, 190], where it is also described how to measure quantities like the quasitemperature, quasi-chemical potentials, and drift velocities, and comparison with experiment is given: theoretical results and experimental data show very good agreement (see also [151, 191, 192]). In that way is made contact with the fundamental point in the scientific method of corroborating theory through comparison with experiment [193]. It is worth mentioning S. J. Gould's observation that "a detail, by itself, is blind; a concept without a concrete illustration is empty [...]. Darwin, who had such keen understanding of fruitfull procedure in science, knew in his guts that theory and observation are Siamese twins, inextricably intertwined and continually interacting" [194]. In particular, in the present question of statistical thermodynamics we restate the call of Riogo Kubo, who expressed that "statistical mechanics has been considered a theoretical endeavour. However, statistical mechanics exists for the sake of the real world, not for fictions. Further progress can only be hoped by close cooperation with experiment" [89].

4.8 A generalized $\mathcal{H}$-theorem

One point that is presently missing is the attempt to extract from this Informational Statistical Thermodynamics (and thus to also provide for a reasonable criterion in phenomenological irreversible thermodynamics) the sign of the entropy production function. This is defined as non-negative in extended irreversible thermodynamics, but such property does not follows immediatly from $P(t)$ of Eq. (4.67). Operator $\mathcal{J}$ on which $\bar{\sigma}$ depends [cf. Eqs. (4.72) and (4.74)] has an extremely complicated expression, even in its alternative form given by Eq. (2.26). There is only one manageable case, the quasi-linear theory of relaxation near equilibrium, when $\bar{\sigma}$ is definite positive, as for example proved in [100].

However, it can be shown that for this informational nonequilibrium statistical thermodynamics there follows a *generalized $\mathcal{H}$-theorem*, in the sense of Jancel [195], which we call a *weak principle of increasing of informational entropy production* (which was briefly mentioned in the previous Chapter, cf. Eq. (3.13)). For that purpose we take into account the definition of the MaxEnt entropy, and resorting, for simplicity, to a classical mechanical description, we can write

$$\bar{S}(t) - \bar{S}(t_0) = -\int d\Gamma \Big[\varrho_\varepsilon(\Gamma|t) \ln \bar{\varrho}(\Gamma|t,0) - \varrho_\varepsilon(\Gamma|t_0) \ln \bar{\varrho}(\Gamma|t_0,0) \Big] \quad (4.82)$$

where Γ is a point in classical phase space, t_0 being the initial time of preparation of the system in the distant past ($t_0 \to -\infty$).

But, because of the initial condition of Eq. (2.16) we have that $\ln \bar{\varrho}(\Gamma|t_0,0) = \ln \varrho_\varepsilon(\Gamma|t_0)$ and, further, since Gibbs entropy, namely

$$S_G(t) = -\int d\Gamma \varrho_\varepsilon(\Gamma|t) \ln \varrho_\varepsilon(\Gamma|t) \quad (4.83)$$

is conserved, that is, it is constant in time [then $S_G(t) = S_G(t_0)$ and we recall that $S_G(t_0) = \bar{S}(t_0,0)$], it follows that

$$\bar{S}(t) - \bar{S}(t_0) = -\int d\Gamma \varrho_\varepsilon(\Gamma|t) [\ln \bar{\varrho}(\Gamma|t,0) - \ln \varrho_\varepsilon(\Gamma|t)] = -\int d\Gamma \varrho_\varepsilon(\Gamma|t) [\mathcal{P}_\varepsilon(t) - 1] \ln \varrho_\varepsilon(\Gamma|t) \ , \quad (4.84)$$

where $\mathcal{P}_\varepsilon$ is the projection operator of Eq. (2.18) and then

$$\bar{S}(t) - \bar{S}(t_0) = \bar{S}(t) - S_G(t) \ . \quad (4.85)$$

Because of the condition of normalization of ϱ_ε and $\bar{\varrho}$, at any time t, we can write Eq. (4.84) in the form

$$\Delta\bar{S}(t) = \bar{S}(t) - \bar{S}(t_0) = \bar{S}(t) - S_G(t) = -\int d\Gamma \varrho_\varepsilon(\Gamma|t) [\ln \bar{\varrho}(\Gamma|t,0) - \ln \varrho_\varepsilon(\Gamma|t)] + \int d\Gamma [\varrho_\varepsilon(\Gamma|t) - \bar{\varrho}(\Gamma|t,0)] \ , \quad (4.86)$$

since the last integral is null. This quantity $\Delta\bar{S}$ cancels for $\varrho_\varepsilon = \bar{\varrho}$ [i.e. for null ϱ'_ε of Eq. (2.17)], and its change when introducing the variation $\varrho_\varepsilon \to \varrho_\varepsilon + \delta\varrho_\varepsilon$ is

$$\delta\Delta\bar{S}(t) = \int d\Gamma \delta\varrho_\varepsilon(\Gamma|t) \ln\left[\frac{\varrho_\varepsilon(\Gamma|t)}{\bar{\varrho}(\Gamma|t,0)}\right] = \\ = \int d\Gamma \delta\varrho_\varepsilon(\Gamma|t) \ln\left[1 + \frac{\varrho'_\varepsilon(\Gamma|t)}{\bar{\varrho}(\Gamma|t,0)}\right] , \quad (4.87)$$

where we used the separation of ϱ_ε as given by Eq. (2.17).

The variation in Eq. (4.87) vanishes for $\varrho_\varepsilon = \bar{\varrho}$, when, as shown, also vanishes $\Delta\bar{S}(t)$, so it follows that $\Delta\bar{S}(t)$ is a minimum for $\varrho_\varepsilon = \bar{\varrho}$, when it is zero and positive otherwise, namely

$$\Delta\bar{S}(t) \geq 0 , \quad (4.88)$$

which defines in MaxEnt-NESOM the equivalent of Jancel's generalized $\mathcal{H}$-theorem [195]. It should be noticed that the inequality of Eq. (4.88) can be interpreted as the fact that, as the system evolves in time from the initial condition of preparation under the governing action of the non-linear generalized kinetic equations the informational entropy cannot decrease, or, because of Eq. (4.88), the informational entropy is at any $t > t_0$ larger (or at most equal in a nondissipative system) than Gibbs statistical entropy. These results reproduce for the MaxEnt-NESOM described in Chapter 2, those obtained by del Rio and García-Colín [167] in an alternative way.

Using the definition for the informational entropy production function we can rewrite Eq. (4.88) in the form of Eq. (3.13), that is

$$\boxed{\Delta\bar{S}(t) = \int_{t_0}^{t} dt' \int d^3r\, \bar{\sigma}(\mathbf{r},t') \geq 0 .} \quad (4.89)$$

Equation (4.89) does not prove that $\bar{\sigma}(\mathbf{r},t)$ is a monotonically increasing function of time, as required by phenomenological irreversible thermodynamics theories. We have only proved the weak condition that as the system evolves $\bar{\sigma}$ is predominantly definite positive, or, better to say, that the informational entropy increases in comparison with its initial value. We also insist on the fact that this result is a consequence of the presence of the contribution ϱ'_ε to ϱ_ε, which is then, as stated previously, the part that accounts for — in the description of the macroscopic state of the system — the processes which generate dissipation. Furthermore, the informational entropy with the evolution property of Eq. (4.88) is the coarse-grained entropy of Eq. (3.2), the coarse-graining being performed by the action of the projection operator $\mathcal{P}_\varepsilon$: This projection operation extracts from the Gibbs entropy the contribution associated to the constraints [cf. Eq. (2.7)] imposed on the system, by projecting it onto the subspace spanned by the basic dynamical quantities, what is graphically illustrated in Fig. 2.1 (see also [181]). Hence, the informational entropy thus defined depends on the choice of the basic set of macroscopic variables, whose completeness in a purely thermodynamic sense cannot be indubitably asserted. We restate that in each particular problem under consideration the information lost as a result of the particular truncation of the set of basic variables must be carefully evaluated [134, 135].

Retaking the question of the signal of $\bar{\sigma}(\boldsymbol{r}, t)$, we conjecture that it is always non-negative, since it can not be understood how information can be gained in some time intervals along the irreversible evolution of the system. However, such property should be expected to be valid as long as one is using a complete description of the system, meaning that the closure condition [Zubarev-Peletminskii rule of Eq. (2.4)] is fully satisfied. Once a truncation procedure is introduced [135], that is, the closure condition is violated, then the local density of informational entropy production is no longer monotonously increasing in time; this has been illustrated by Criado-Sancho and Llebot [196] in the realm of Extended Irreversible Thermodynamics, and in IST in [197]. The reason is, as pointed out by Balian et al. [181] that the truncation procedure introduces some kind of additional (spurious) information at the step when the said truncation is imposed [in the illustration given in section 4.5, this occurs when the second order flux is written in terms of the reduced set of basic variables, namely the density of energy and its fluxes, leading to Eqs. (4.58)].

Two other properties of the MaxEnt entropy function are that, first, it is a maximum compatible with the constraints of Eq. (2.7) when they are given at the specific time t, that is, $\bar{\varrho}$ maximizes $\bar{S}(t)$ when subjected to normalization and such constraints. This particular property of entailed maximization, which ensures that $\bar{S}(t)$ is a convex function in the space of thermodynamic states, is the one that concomitantly ensures that in the framework of Informational Statistical Thermodynamics are contained generalized forms of Prigogine's theorem of minimum entropy production in the linear regime around equilibrium, and Glansdorff-Prigogine's thermodynamic principles of evolution and (in)stability in nonlinear conditions. Let us see these points next.

4.9 Evolution and (in)stability criteria

In this section we summarize three additional properties of the informational-entropy production, namely the criterion of evolution and the (in)stability criterion — generalizations of those of Glansdorff-Prigogine in nonlinear classical Nonequilibrium Thermodynamics [20], and a criterion for minimum production of informational entropy also a generalization of the one due to Prigogine [17]. Details of the demonstrations are given elsewhere [166].

The time derivative of $\bar{\sigma}(t)$ of Eq. (3.10) can be split into two terms, namely

$$\frac{d\bar{\sigma}(t)}{dt} = \frac{d_F\bar{\sigma}(t)}{dt} + \frac{d_Q\bar{\sigma}(t)}{dt} , \tag{4.90}$$

where

$$\frac{d_F\bar{\sigma}(t)}{dt} = \sum_{j=1}^{n} \int d^3r \frac{\partial F_j(\mathbf{r},t)}{\partial t} \frac{\partial Q_j(\mathbf{r},t)}{\partial t} , \tag{4.91}$$

and

$$\frac{d_Q\bar{\sigma}(t)}{dt} = \sum_{j=1}^{n} \int d^3r F_j(\mathbf{r},t) \frac{\partial^2 Q_j(\mathbf{r},t)}{\partial t^2} , \tag{4.92}$$

that is, the change in time of the informational-entropy production due to that of the variables F and that of the variables Q, respectively. Recalling that the time-derivative of Q_j stand for the thermodynamic variables in Informational Statistical Thermodynamics [cf. Eq. (2.6)], while the F_j stand for the Lagrange multipliers introduced by the method [cf. Eqs. (3.9)], which are the differential coefficients of the informational entropy [cf. Eq. (4.48)], Eqs. (4.91) and (4.92) are related to what, in the limiting case of classical (linear or Onsagerian) Thermodynamics, are the contributions due to the change in time of the thermodynamic fluxes and forces respectively, as will be better clarified in continuation.

First we notice that using the fact that $F_j(\boldsymbol{r},t) = \delta\bar{S}(t)/\delta Q_j(\boldsymbol{r},t)$, we can write the equations of evolution for the $\{F_j\}$ in terms of those for the $\{Q_j\}$, namely

$$\frac{\partial F_j(\boldsymbol{r},t)}{\partial t} = \frac{\partial}{\partial t}\frac{\delta\bar{S}(t)}{\delta Q_j(\boldsymbol{r},t)} = \\ = \sum_{k=1}^{n}\int d^3r_1 \frac{\delta^2\bar{S}(t)}{\delta Q_j(\boldsymbol{r},t)\delta Q_k(\boldsymbol{r}_1,t)}\frac{\partial Q_k(\boldsymbol{r}_1,t)}{\partial t}, \tag{4.93}$$

and then

$$\frac{d_F\bar{\sigma}(t)}{dt} = \sum_{j,k=1}^{n}\int d^3r\int d^3r_1 \frac{\delta^2\bar{S}(t)}{\delta Q_j(\boldsymbol{r},t)\delta Q_k(\boldsymbol{r}_1,t)}\frac{\partial Q_j(\boldsymbol{r},t)}{\partial t}\frac{\partial Q_k(\boldsymbol{r}_1,t)}{\partial t}, \tag{4.94}$$

which is non-positive because of the convexity of $\bar{S}$ in the space spanned by variables $\{Q_j\}$, as shown in [166], that is

$$\boxed{\frac{d_F\bar{\sigma}(t)}{dt} \leq 0\,.} \tag{4.95}$$

Inequality (4.95) can be considered as a *generalized Glansdorff-Prigogine thermodynamic criterion for evolution,* which in our approach is a consequence of the use of MaxEnt in the construction of IST within the framework of Predictive Statistical Mechanics.

Also, an alternative criterion can be derived in terms of the generating functional $\phi(t)$, as given by Zubarev [100]. Defining

$$\varphi(t) = \frac{d\phi(t)}{dt} = -\sum_{j=1}^{n}\int d^3r Q_j(\boldsymbol{r},t)\frac{\partial F_j(\boldsymbol{r},t)}{\partial t}, \tag{4.96}$$

it follows that

$$\frac{d\varphi(t)}{dt} = \frac{d_F\varphi(t)}{dt} + \frac{d_Q\varphi(t)}{dt}, \tag{4.97}$$

where

$$\frac{d_F\varphi(t)}{dt} = -\sum_{j=1}^{n}\int d^3r Q_j(\mathbf{r},t)\frac{\partial^2 F_j(\mathbf{r},t)}{\partial t^2}\,, \tag{4.98}$$

and

$$\frac{d_Q\varphi(t)}{dt} = -\sum_{j=1}^{n}\int d^3r \frac{\partial Q_j(\mathbf{r},t)}{\partial t}\frac{\partial F_j(\mathbf{r},t)}{\partial t}\,. \tag{4.99}$$

But $d_Q\varphi(t)/dt$ is then $-d_F\bar{\sigma}(t)/dt$, and because of Eq. (4.95)

$$\boxed{\frac{d_Q\varphi(t)}{dt} \geq 0} \tag{4.100}$$

during the irreversible evolution of the system, hence it follows an alternative criterion given in terms of the variation in time of the rate of change of the logarithm of the nonequilibrium partition-like function.

Consider next an isolated system composed of a given open system in interaction with the rest acting as sources and reservoirs. These sources and reservoirs are assumed to be ideal, that is, their statistical distributions, denoted by $\varrho_\Sigma\varrho_R$, are taken as constantly stationary, in order words as unaltered by the interaction with the much smaller open system. The auxiliary ("coarse-grained" or "instantaneously frozen") nonequilibrium statistical operator, $\bar{\varrho}(t,0)$, for the whole system to be used in the equation of evolution is then written as

$$\bar{\varrho}(t,0) = \bar{\varrho}_S(t,0) \times \varrho_\Sigma\varrho_R\,, \tag{4.101}$$

where $\bar{\varrho}_S(t,0)$ is the auxiliary statistical operator of the open system constructed in the MaxEnt-NESOM, and we recall that the nonequilibrium statistical operator, $\varrho_\varepsilon(t)$, is built in terms of this auxiliary $\bar{\varrho}(t,0)$, as indicated by Eq. (2.12) [cf. section 4.2].

If the open system is in a steady state [to be denoted hereafter by an upper index (ss)], then the production of global informational entropy is null, that is, $\bar{\sigma}^{(ss)} = \bar{\sigma}^{(ss)}_{(i)} + \bar{\sigma}^{(ss)}_{(e)} = 0$, where the two contributions are the internal and external production of informational entropy, as given by Eqs. (4.75) and (4.76). Hence, $\bar{\sigma}^{(ss)}_{(i)} = -\bar{\sigma}^{(ss)}_{(e)}$, meaning that the increase of the global internal entropy production is compensated by the pumping of entropy to the external world.

Let us consider now a small deviation from the steady state which is assumed to be *near equilibrium,* and we write

$$F_j(\boldsymbol{r},t) = F_j^{(\text{eq})} + \Delta F_j(\boldsymbol{r},t) \,, \tag{4.102}$$

where $\Delta F \ll F^{(\text{eq})}$, and index (eq) indicates the value of the corresponding quantity in the *equilibrium state.* The internal production of informational entropy, $\bar{\sigma}_{(i)} = P(t)$ of Eqs. (4.75) and (4.69), in the condition of departure from equilibrium defined by Eq. (4.102), satisfies in this *immediate neighborhood of the steady state near equilibrium,* which we call the *strictly linear regime* (SLR), that

$$\bar{\sigma}_{(i)}^{\text{SLR}}(t) = P^{\text{SLR}}(t) = \sum_{j,k=1}^{n} \int d^3r \int d^3r_1 \Delta F_j(\boldsymbol{r},t) L_{jk}^{\text{SLR}}(\boldsymbol{r},\boldsymbol{r}_1) \Delta F_k(\boldsymbol{r}_1,t) + $$
$$+ \sum_{j=1}^{n} \int d^3r \nabla \Delta F_j(\boldsymbol{r},t) \otimes \tilde{I}_j(\boldsymbol{r},t) \,, \tag{4.103}$$

for generic $\{Q_j\}$ and $\{F_j\}$, which, we recall, in the case of the nonequilibrium grand-canonical ensemble corresponds to the two densities of energy and particles and their fluxes of all order, and the corresponding Lagrange multipliers, respectively [cf. Eqs. (4.30) and (4.28)]. Moreover, $\tilde{I}$ stands for all type of fluxes, say the sets $\{I_n^{[r]}\}$ and $\{I_h^{[r]}\}$, and the products of the quantities involved are to be understood as algebraic, scalar, and fully contracted tensorial as the case demands. Indexes j in the case of the generalized grand-canonical are then $j = 1$ for h, $j = 2$ for n, $j = 3$ for $\boldsymbol{I}_h$, $j = 4$ for $\boldsymbol{I}_n$, $j = 2m+1$ for $I_h^{[r]}$ and $j = 2m$ for $I_n^{[r]}$, with $m = 2, 3, \ldots$ and $m = r$. In Eq. (4.103) L^{SLR} is an Onsager-like symmetric matrix of kinetic coefficients, around the equilibrium state, given by

$$L_{jk}^{\text{SLR}}(\boldsymbol{r},\boldsymbol{r}_1) = \int_{-\infty}^{0} d\tau e^{\varepsilon\tau} [\text{Tr}\{\dot{\hat{P}}'_j(\boldsymbol{r}_1)\dot{\hat{P}}'_k(\boldsymbol{r},\tau)\varrho_{eq}\} - $$
$$- \sum_{m,n} \text{Tr}\{\hat{P}_j(\boldsymbol{r})\dot{\hat{P}}'_m(\boldsymbol{r}_1) C_{mn}^{-1}(\boldsymbol{r},\boldsymbol{r}_1)\hat{P}_n(\boldsymbol{r})\dot{\hat{P}}'_k(\boldsymbol{r}_1)\varrho_{eq}\}] \,, \tag{4.104a}$$

where

$$C_{mn}(\boldsymbol{r},\boldsymbol{r}_1) = \int_0^1 du \; \text{Tr}\{\hat{P}_m(\boldsymbol{r})\varrho_{eq}^u \hat{P}_n(\boldsymbol{r}_1)\varrho_{eq}^{-u+1}\} \,, \tag{4.105}$$

and

$$\dot{\hat{P}}' = \frac{1}{i\hbar}[\hat{P},\hat{H}'] , \tag{4.106}$$

with $\hat{H}'$ of Eq. (2.3), and use was made of the fact that

$$\sum_j \int d^3r \int d^3r_1 L_{jk}^{\mathrm{SLR}}(\boldsymbol{r},\boldsymbol{r}_1)F_j^{(eq)} = \sum_k \int d^3r \int d^3r_1 L_{jk}^{\mathrm{SLR}}(\boldsymbol{r},\boldsymbol{r}_1)F_k^{(eq)} = 0 , \tag{4.107}$$

i.e., there is no production of informational entropy in equilibrium. Moreover, in the neighborhood of the equilibrium state the internal production of entropy is nonnegative, that is (see for example [100])

$$\bar{\sigma}_{(i)}^{\mathrm{SLR}}(t) = P^{\mathrm{SLR}}(t) \geq 0 . \tag{4.108}$$

Taking into account that for a system subject to time-independent external constraints, so as to produce a steady state, it is verified that $dP_\Sigma/dt = 0$, we obtain that

$$\frac{d_F\bar{\sigma}_{(i)}^{\mathrm{SLR}}(t)}{dt} = \sum_{j,k=1}^{n} \int d^3r \int d^3r_1 \frac{\partial \Delta F_j(\boldsymbol{r},t)}{\partial t} L_{jk}^{\mathrm{SLR}}(\boldsymbol{r},\boldsymbol{r}_1)\Delta F_k(\boldsymbol{r}_1,t) + \sum_{j=1}^{n} \int d^3r \nabla \frac{\partial \Delta F_j(\boldsymbol{r},t)}{\partial t} \otimes \tilde{I}_j(\boldsymbol{r},t) , \tag{4.109}$$

and

$$\frac{d_Q\bar{\sigma}_{(i)}^{\mathrm{SLR}}(t)}{dt} = \sum_{j,k=1}^{n} \int d^3r \int d^3r_1 \Delta F_j(\boldsymbol{r},t) L_{jk}^{\mathrm{SLR}}(\boldsymbol{r},\boldsymbol{r}_1) \frac{\partial \Delta F_k(\boldsymbol{r}_1,t)}{\partial t} + + \sum_{j=1}^{n} \int d^3r \nabla \Delta F_j(\boldsymbol{r},t) \otimes \frac{\partial \tilde{I}_j(\boldsymbol{r},t)}{\partial t} . \tag{4.110}$$

But, according to Onsager's relations in the linear domain around equilibrium (see for example [100]), it follows that

$$\tilde{I}_j(\boldsymbol{r},t) = \sum_{j=1}^{n} \int d^3r_1 \Lambda_{jk}^{\mathrm{SLR}}(\boldsymbol{r},\boldsymbol{r}_1)\, \nabla \Delta F_k(\boldsymbol{r}_1,t) , \tag{4.111}$$

where $\Lambda^{\text{SLR}}_{jk}(\boldsymbol{r},\boldsymbol{r}_1) = \Lambda^{\text{SLR}}_{kj}(\boldsymbol{r},\boldsymbol{r}_1)$. i.e., the matrix of kinetic coefficients is symmetric. Using Eq. (4.111) and recalling that the matrix of coefficients L^{SLR}_{jk} is also symmetric [166], we can verify that the expressions in Eqs. (4.109) and (4.110) are identical and, therefore,

$$\frac{d\bar{\sigma}^{\text{SLR}}_{(i)}(t)}{dt} = (\frac{d_F}{dt} + \frac{d_Q}{dt})\bar{\sigma}^{\text{SLR}}_{(i)}(t) = 2\frac{d_F\bar{\sigma}^{\text{SLR}}_{(i)}(t)}{dt} . \tag{4.112}$$

The results of Eqs. (4.75), (4.76), (4.107) and (4.112) are of relevance in proving a generalization in IST of Prigogine's theorem of minimum internal production of entropy. In fact, in the SLR regime, taking the time derivative of Eq. (4.67), it follows from Eq. (4.112) that

$$\frac{dP^{\text{SLR}}}{dt} = \frac{d\bar{\sigma}^{\text{SLR}}_{(i)}}{dt} = 2\frac{d_F\bar{\sigma}^{\text{SLR}}_{(i)}}{dt} = 2\frac{d_F\bar{\sigma}^{\text{SLR}}}{dt} \le 0 , \tag{4.113}$$

as a consequence that in the steady state the fluxes $\tilde{I}_j$ in Eq. (4.103) are time independent on the boundaries, i.e. $dP_\Sigma/dt = 0$, as already noticed. Hence, the inequality in Eq. (4.113) is a consequence for this particular case of the theorem of Eq. (4.95). Therefore, on account of Eqs. (4.108) and (4.113), according to Lyapunov's theorem (e.g. see for example Ref. [169]) there follows the generalization in IST of Prigogine's theorem.

This theorem proves that in the linear regime near equilibrium, that is in the strictly linear regime, $\bar{\sigma}^{\text{SLR}}_{(i)}$ is a nonequilibrium state function with an associated variational principle, and that *steady states near equilibrium are attractors characterized by producing the least dissipation (least loss of information in our theory) under the given constraints.* We stress that this result is a consequence of the fact that the matrix of kinetic coefficients L_{jk} is *symmetric in the strictly linear regime* and this result implies that $d_F\bar{\sigma}^{\text{SRL}}_{(i)}$ near equilibrium is one half the exact differential of the entropy production. *Outside the strictly linear regime an antisymmetric part of the matrix of kinetic coefficients may be present and, therefore, there is no variational principle that could ensure the stability of the steady state.* The latter may become unstable once a certain distance from equilibrium is attained, giving rise to the emergence of a selforganized dissipative structure in Prigogine's sense [84, 87, 169, 198].

Consequently, outside the strictly linear domain, essentialy for systems far-away-from equilibrium, other stability criterion needs be determined. We look for it first noting that, because of the convexity of the MaxEnt-NESOM entropy in the space of basic macrovariables $\{Q_j\}$, as is demonstrated in [166] for states around any arbitrary steady state, let it be near or far away from equilibrium, the following result, for generic $\{Q_j\}$ and $\{F_j\}$, holds, namely

$$\frac{1}{2}\delta^2\bar{S}(t) = \frac{1}{2}\sum_{j,k=1}^{n}\int d^3r\int d^3r_1 C_{jk}^{(-1)}(\boldsymbol{r},\boldsymbol{r}_1)^{(ss)}\Delta Q_j(\boldsymbol{r},t)\Delta Q_k(\boldsymbol{r}_1,t) \leq 0\,, \tag{4.114}$$

where

$$C_{jk}^{(-1)}(\boldsymbol{r},\boldsymbol{r}_1)^{(ss)} = \left[\frac{\delta^2\bar{S}(t)}{\delta Q_j(\boldsymbol{r},t)\delta Q_k(\boldsymbol{r}_1,t)}\right]^{(ss)}\,, \tag{4.115}$$

meaning that the value of the second order variational derivative is taken in the steady state.

Time derivation of Eq. (4.114) leads to the expression

$$\begin{aligned}\Delta_F\bar{\sigma}(t) &\equiv \frac{d}{dt}\frac{1}{2}\delta^2\bar{S}(t) = \\ &= \sum_{j,k=1}^{n}\int d^3r\int d^3r_1 C_{jk}^{(-1)}(\boldsymbol{r},\boldsymbol{r}_1)^{(ss)}\Delta Q_k(\boldsymbol{r},t)\frac{\partial\Delta Q_j(\boldsymbol{r}_1,t)}{\partial t} = \\ &= \sum_{j=1}^{n}\int d^3r\Delta F_j(\boldsymbol{r},t)\frac{\partial\Delta Q_j(\boldsymbol{r},t)}{\partial t} = \\ &= \bar{\sigma}(t) - \sum_{j=1}^{n}\int d^3r\, F_j^{(ss)}(\boldsymbol{r},t)\frac{\partial\Delta Q_j(\boldsymbol{r},t)}{\partial t}\,,\end{aligned} \tag{4.116}$$

a quantity called the *excess entropy production.* In deriving the previous to the last term in Eq. (4.116) we used in the calculation that $C^{(-1)}$, the inverse of matrix C, is taken in the steady state and thus is time independent, and it was used the fact that C is a symmetric matrix (what is a manifestation of the existence of Maxwell-type relations in IST, as will be shown later on) [see Section 4.11].

Consequently, taking into account Eqs. (4.114) and (4.116), Lyapunov stability theorem (see for example [169]) allows us to establish a generalized *Glansdorff-Prigogine-like (in)stability criterion* in the realm of Informational Statistical Thermodynamics:

For given constraints, if $\Delta_F\sigma$ is positive, the reference steady state is stable for all fluctuations compatible with the equation of evolution (which are provided in MaxEnt-NESOM by the nonlinear transport theory briefly summarized in Chapter 2).

Stability of the equilibrium and steady states in the linear regime around equilibrium are recovered as particular limiting and restricted cases of the general theory. Therefore, given a dynamical open system in a certain steady state, it can be driven away from it by changing one or more control parameters on which its macrostate depends. At some critical value of one or more of these control parameters (for example the intensities of external fields) the sign of the excess entropy production function may change from positive to negative, meaning that the steady state loses its stability and a new macrostate becomes stabilized characterizing some kind of, in general, patterned structure, the so-called *Prigogine's dissipative structure.*

The character of the emerging structure is connected with the type of fluctuation arising in the system that instead of regressing, as should be the case in the linear (or Onsagerian) regime, increases to create the new macrostate. The instability corresponds to a branching point of solutions of the non-linear equations of evolution, and to maintain such non-equilibrium structures a continous exchange of energy and/or matter with external reservoirs is necessary, i.e. entropy must be pumped out of the open system. We again stress that this kind of self-organized ordered behavior is ruled out in the strictly linear regime as a result of the previously demonstrated generalized Prigogine's minimum entropy production theorem. Thus, nonlinearity is required for these structural transitions to occur at a sufficiently far distance from equilibrium. The new nonequilibrium branch of solution (the dissipative structure) may present one of the following three characteristics [87, 169]:

(a) time organization,

(b) space organization,

(c) multiple steady states.

Finally, we stress the fact that in the nonlinear domain the criteria for evolution and stability are decoupled — differently to the linear domain — and this fact allows for the occurrence of new types of behavior when the dynamical system is driven far away from equilibrium: order may arise out of thermal chaos [127, 199] and the system displays complex behavior. As noted before, and we restate here, Informational Statistical Thermodynamics, built within the framework of MaxEnt-NESOM, can provide solid microscopic and macroscopic basis to deal with selforganization, or the sometimes called Thermodynamics of Complex Systems [200].

4.10 Generalized Clausius relation in IST

The informational entropy in IST also satisfies a kind of generalized Clausius relation. In fact, consider the modification of the informational entropy as a consequence of the modification of external constraints imposed on the system. Let us call λ_l $(l = 1, 2, \ldots, s)$ a set of parameters that characterize such constraints (e.g., the volume, particle number, external fields, etc.). Introducing infinitesimal modifications of them, say $d\lambda_l$, the corresponding variation in the informational entropy is, for generic $\{Q_j\}$ and $\{F_j\}$, given by

$$\boxed{d\bar{S}(t) = \int d^3r \sum_{j=1}^{n} F_j(\boldsymbol{r},t)\delta Q_j(\boldsymbol{r},t)\,,} \tag{4.117}$$

where δQ_j are the *nonexact differentials*

$$\delta Q_j(\boldsymbol{r},t) = dQ_j(\boldsymbol{r},t) - \langle d\hat{P}_j(\boldsymbol{r})|t\rangle\,, \tag{4.118}$$

with $\langle d\hat{P}_j(\boldsymbol{r})|t\rangle = \mathrm{Tr}\{d\hat{P}_j\bar{\varrho}(t,0)\}$. In these expressions the nonexact differentials are the difference between the exact differential

$$dQ_j(\boldsymbol{r},t) = d\,\mathrm{Tr}\Big\{\hat{P}_j(\boldsymbol{r})\bar{\varrho}(t,0)\Big\} = \sum_{l=1}^{s} \frac{\partial Q_j(\boldsymbol{r},t)}{\partial \lambda_l} d\lambda_l\,, \tag{4.119}$$

with

$$\langle d\hat{P}_j(\boldsymbol{r})|t\rangle = \mathrm{Tr}\Big\{\sum_{l=1}^{s} \frac{\partial \hat{P}_j(\boldsymbol{r})}{\partial \lambda_l} d\lambda_l\,\bar{\varrho}(t,0)\Big\}\,, \tag{4.120}$$

the latter being the average value of the change in the corresponding dynamical quantity due to the modification of the control parameters. This follows from the fact that

$$d\bar{S}(t) = \sum_{l=1}^{s} \frac{\partial \phi(t)}{\partial \lambda_l} + \sum_{j=1}^{n} \sum_{l=1}^{s} \int d^3r \Big[\frac{\partial F_j(\mathbf{r},t)}{\partial \lambda_l} Q_j(\mathbf{r},t) + F_j(\mathbf{r},t) \frac{\partial Q_j(\mathbf{r},t)}{\partial \lambda_l} d\lambda_l \Big] , \tag{4.121}$$

and that

$$\frac{\partial \phi(t)}{\partial \lambda_l} = \frac{\partial}{\partial \lambda_l} \mathrm{Tr}\Big\{ \exp\Big[- \sum_{j=1}^{n} \int d^3r F_j(\mathbf{r},t) \hat{P}_j(\mathbf{r}) \Big] \Big\} = - \sum_{j=1}^{n} \int d^3r \left[Q_j(\mathbf{r},t) \frac{\partial F_j(\mathbf{r},t)}{\partial \lambda_l} + F_j(\mathbf{r},t) \,\mathrm{Tr}\Big\{ \frac{\partial \hat{P}_j(\mathbf{r})}{\partial \lambda_l} \bar{\varrho}(t,0) \Big\} \right] . \tag{4.122}$$

Consequently, using Eqs. (4.121) and (4.122), we obtain Eq. (4.117). It is worth noticing that if we take the system in equilibrium at temperature T, described by the canonical distribution, and perform an infinitesimal change in volume, say dV, then

$$dS = \frac{1}{T} \left[d\langle \hat{H} \rangle - \langle d\hat{H} \rangle \right] = \frac{dU}{T} - \frac{1}{T} \langle \frac{\partial \hat{H}}{\partial V} dV \rangle = \frac{dU}{T} + \frac{p}{T} dV \tag{4.123}$$

where p is the pressure given by $p = - \langle \partial \hat{H} / \partial V \rangle$, and then follows the form of the first law given by

$$dU = TdS - pdV . \tag{4.124}$$

Equation (4.117) tells us that the MaxEnt-NESOM Lagrange multipliers are *integrating factors* for the nonexact differentials δQ_j.

Let us take as the energy density one of the variables Q_j, say $Q_1(\mathbf{r},t) = \epsilon(\mathbf{r},t)$, and in analogy with equilibrium we define the intensive non-equilibrium thermodynamic variable we call *quasitemperature*, or better to say its reciprocal [cf. Eq. (4.80)]

$$\Theta^{-1}(\mathbf{r},t) = \delta \bar{S}(t) / \delta \epsilon(\mathbf{r},t) . \tag{4.125}$$

After introducing the additional redefinitions of the Lagrange multipliers in the form

$$F_j(\mathbf{r},t) = \Theta^{-1}(\mathbf{r},t) \, \mathcal{F}_j(\mathbf{r},t) , \tag{4.126}$$

using Eq. (3.4) allows us to introduce a kind of generalized space-dependent Clausius expression for a system in arbitrary nonequilibrium conditions, namely

$$\bar{\eta}(\boldsymbol{r},t) - \bar{\eta}(\boldsymbol{r},t_0) = \int_{t_0}^{t} dt' \frac{\partial \eta(\boldsymbol{r},t')}{\partial t'} = \int_{t_0}^{t} dt' \Theta^{-1}(\boldsymbol{r},t')\, \delta\dot{q}(\boldsymbol{r},t')\,, \quad (4.127)$$

where we have introduced the nonexact differential for a *generalized heat function* $q(\boldsymbol{r},t)$ given by

$$\delta q(\boldsymbol{r},t') = dt'\,\delta\dot{q}(\boldsymbol{r},t') = \delta\epsilon(\boldsymbol{r},t') + \sum_{j=2}^{n} \mathcal{F}_j(\boldsymbol{r},'t)\,\delta Q_j(\boldsymbol{r},t')\,. \quad (4.128)$$

In Eq. (4.127) is to be understood that the integration in time extends along the trajectory of evolution of the system, governed by the kinetic equations (2.24) [or (4.31)].

Moreover, using the redefinitions given in Eqs. (4.126), we may noticed that the generalized space and time dependent Gibbs relation of Eq. (3.4) becomes

$$\boxed{\Theta(\boldsymbol{r},t)d\eta(\boldsymbol{r},t) = d\epsilon(\boldsymbol{r},t) + \sum_{j=2}^{n} \mathcal{F}_j(\boldsymbol{r},t)\,dQ_j(\boldsymbol{r},t)\,,} \quad (4.129)$$

where on the left side it has been put into evidence the quasitemperature Θ (we stress that in Eq. (4.128) δQ indicates the nonexact differential resulting from modifications in the constraints, while in Eq. (4.129) is present the differential dQ of the variables on which the function $\bar{S}$ depends).

This is the case of a single nonequilibrium system; we have already call the attention to the fact that, in the general case, there exist different quasitemperatures for different subsets of degrees of freedom of a given sample in nonequilibrium conditions. The case of the photoinjected highly excited plasma in semiconductors is an excellent example: Coulomb interaction between carriers produces their internal thermalization (in nonequilibrium conditions) resulting in a unique quasitemperature which is usually attained in the ten-fold femtosecond time scale, while the optical phonons are driven away from equilibrium as a result of the interaction with the nonequilibrium carriers and acquire different quasitemperatures in each mode. Only in the ten-fold picosecond time scale there follows mutual thermalization of carriers and all the phonon modes, when all acquire a unique quasitemperature, and in the long run carriers' and optical-phonons' systems attain final equilibrium with the heat reservoir with which they are in contact; then the quasitemperature goes over the temperature of equilibrium with the reservoir [151, 201].

Integrating in space Eq. (4.128), and taking into account the results that led to the $\mathcal{H}$-theorem of section 2.3, we can write

$$\bar{S}(t) - \bar{S}(t_0) = \bar{S}(t) - S_G(t) = \int_{t_0}^{t} dt' \int d^3r \Theta^{-1}(\mathbf{r}, t')\delta\dot{q}(\mathbf{r}, t') \geq 0 \,. \tag{4.130}$$

Inspection of this Eq. (4.130) suggests a "free interpretation" of it as resulting from a kind of pseudo-Carnot principle for arbitrary nonequilibrium systems, in the sense of taking the contribution of the integrand as a local reversible exchange of a heat-like quantity between the system and a pseudo-reservoir at local and instantaneous temperature $\Theta(\mathbf{r}, t)$. Some considerations on Carnot's principle and its connection with MaxEnt, as a general principle of reasoning, has been advanced by Jaynes [202]. He described the evolution of Carnot's principle, via Kelvin's perception that it defines a univeral temperature scale, Clausius' discovery that it implied the existence of the entropy function, Gibbs' perception of its logical status, and Boltzmann's interpretation of entropy in terms of phase volume into the general formalism of statistical mechanics. The equivalent in MaxEnt of Boltzmann's results is provided in section 4.12.

At this point we may comment that the approach to irreversible thermodynamics provided by IST appears to have a certain relationship with we have called Orthodox Irreversible Thermodynamics carried on by Chan-Eu [24] which has a general appealing outlook. His state function "calortropy" has a certain similarity with our IST-informational statistical entropy. However, apparently, there are differences, one is on the physical interpretation and definition of the space of state variables, and mainly on the question of a nonequilibrium temperature. Other is the question, as some authors have pointed out, of the operativeness of the method, which has not been clearly demonstrated. In particular how to deal with actual physical situations, that is, in real experimental conditions, what in IST has been done somewhat extensively and successfully as noticed in several references along the text. This has involved the cases of semiconductors, polymers, and biopolymers, what is being validating the mechano-statistical approach in IST.

4.11 Fluctuations and Maxwell-like relations

As already shown, the average value of any dynamical quantity $P_j(\Gamma)$ (in the case of the nonequilibrium grand-canonical ensemble the two densities of energy and particles and their fluxes of all order) of the basic set in MaxEnt-NESOM (the classical mechanical level of description is used for simplicity) is given by

$$Q_j(t) = \int d\Gamma P_j(\Gamma)\; \bar{\varrho}(\Gamma|t,0) = -\delta\phi(t)/\delta F_j(t)\,, \tag{4.131}$$

that is, by minus the functional derivative of the generating functional ϕ with respect to the associated Lagrange multiplier $F_j(t)$ [and we recall that this function can be related to a kind of non-equilibrium partition function through the expression $\phi(t) = \ln \bar{Z}(t)$]. Moreover, from a straight calculation it follows that

$$\begin{aligned} \delta^2\phi(t)/\delta F_j(t)\delta F_k(t) &= -\delta Q_j(t)/\, \delta F_k(t) = -\delta Q_k(t)/\, \delta F_j(t) = \\ &= \int d\Gamma \Delta P_j(\Gamma)\Delta P_k(\Gamma)\bar{\varrho}(\Gamma|t,0) = C_{jk}(t), \end{aligned} \tag{4.132}$$

where

$$\Delta P_j(\Gamma) = P_j(\Gamma) - \int d\Gamma P_j(\Gamma)\bar{\varrho}(\Gamma|t,0) = P_j(\Gamma) - Q_j(t)\,, \tag{4.133}$$

and Eq. (4.132) defines the *matrix of correlations* $\hat{C}(t)$. The diagonal elements of $\hat{C}$ are the mean square deviations, or fluctuations, of quantities $P_j(\Gamma)$, namely

$$\begin{aligned} C_{jj}(t) &= \int d\Gamma \left[\Delta P_j(\Gamma)\right]^2 \bar{\varrho}(\Gamma|t,0) = \\ &= \int d\Gamma \left[P_j(\Gamma) - Q_j(t)\right]^2 \bar{\varrho}(\Gamma|t,0) \equiv \Delta^2 Q_j(t)\,, \end{aligned} \tag{4.134}$$

and the matrix is symmetrical, that is,

$$\boxed{C_{jk}(t) = \frac{\delta^2\phi(t)}{\delta F_j(t)\delta F_k(t)} = \frac{\delta^2\phi(t)}{\delta F_k(t)\delta F_j(t)} = C_{kj}(t)\,,} \tag{4.135}$$

what is a manifestation in IST of the known *Maxwell relations* in equilibrium.

Let us next scale the informational entropy and Lagrange multipliers in terms of Boltzmann constant, k_B, that is, we introduce

$$S(t) = k_B \bar{S}(t); \qquad \mathcal{F}_j(t) = k_B F_j(t) , \tag{4.136}$$

and then, because of Eq. (3.9),

$$\mathcal{F}_j(t) = \delta S(t) / \delta Q_j(t) . \tag{4.137}$$

Moreover, we find that

$$\frac{\delta^2 S(t)}{\delta Q_j(t) \delta Q_k(t)} = \frac{\delta \mathcal{F}_j(t)}{\delta Q_k(t)} = \frac{\delta \mathcal{F}_k(t)}{\delta Q_j(t)} = -k_B C_{jk}^{(-1)}(t) , \tag{4.138}$$

that is, the second order functional derivatives of the MaxEnt-NESOM entropy are the components of minus the inverse of the matrix of correlations $C^{(-1)}$, with elements to be denoted by $C_{jk}^{(-1)}(t)$. Moreover, the fluctuation of the MaxEnt-NESOM entropy is given by

$$\Delta^2 S(t) = \sum_{jk} \frac{\delta S(t)}{\delta Q_j(t)} \frac{\delta S(t)}{\delta Q_k(t)} C_{jk}(t) = \sum_{jk} C_{jk}(t) \mathcal{F}_j(t) \mathcal{F}_k(t) , \tag{4.139}$$

and that of the intensive variables $\mathcal{F}_j$ are

$$\begin{aligned} \Delta^2 \mathcal{F}_j(t) &= \sum_{kl} \frac{\partial \mathcal{F}_j(t)}{\delta Q_k(t)} \frac{\partial \mathcal{F}_j(t)}{\delta Q_l(t)} C_{jk}(t) \\ &= k_B^2 \sum_{kl} C_{jk}^{(-1)}(t) C_{jl}^{(-1)}(t) C_{kl}(t) = k_B^2 C_{jj}^{(-1)}(t) , \end{aligned} \tag{4.140}$$

therefore

$$\Delta^2 Q_j(t) \Delta^2 \mathcal{F}_j(t) = k_B^2 C_{jj}(t) \, C_{jj}^{(-1)}(t) = k_B^2 G_{jj}(t) , \tag{4.141}$$

where

$$G_{jj}(t) = C_{jj}(t) C_{jj}^{(-1)}(t) \tag{4.142}$$

and then

$$\left[\Delta^2 Q_j(t)\right]^{1/2} \left[\Delta^2 \mathcal{F}_j(t)\right]^{1/2} = k_B G_{jj}^{1/2}(t) . \tag{4.143}$$

Recalling that the quantities $C_{jk}^{(-1)}$ are the matrix elements of the inverse of the matrix of correlations, if the variables are uncorrelated then $G_{jj}(t) = 1$. Equation (4.143) has the likeness of an uncertainty principle connecting the variables Q_j and $\mathcal{F}_j(t)$, which are thermodynamically conjugated in the sense of Eqs. (3.9), with Boltzmann constant being the atomistic parameter playing a role resembling that of the quantum of action in mechanics. This leads to the possibility to relate the results of IST with the idea of complementarity between the microscopic and macroscopic descriptions of many-body systems advanced by Rosenfeld and Prigogine [125, 203–205]; this is discussed elsewhere [206] and revisited in section 4.13.

Care must be excercised in referring to fluctuations of the intensive variables F_j. In the statistical description fluctuations are associated to the specific variables Q_j, but the F's are Lagrange multipliers fixed by the average values of the P's, and so $\Delta^2 \mathcal{F}$ is not a proper fluctuation of $\mathcal{F}$ but a second order deviation interpreted as being a result of the fluctuations of the variables Q on which it depends, in a generalization of the usual results in statistical mechanics in equilibrium [207]. These brief considerations point to the desirability to develop a complete theory of fluctuations in the context of MaxEnt-NESOM; one relevant application of it would be the study of the kinetics of transition between dissipative structures in complex systems, of which is presently available a phenomenological approach [169].

4.12 A Boltzmann-like relation

According to the results of the previous subsection, quite similarly to the case of equilibrium it follows that the quotient between the root mean square deviation of a given quantity and its average value is of the order of the reciprocal of the square root of the number of particles, that is

$$\left[\Delta^2 Q_j(t)\right]^{1/2} / Q_j(t) \sim N^{-1/2} \,. \tag{4.144}$$

Consequently, again quite in analogy with the case of equilibrium, the number of states contributing for the quantity P_j to have the given average value, is overhelmingly enormous (a rigorous demonstration follows resorting to the method of the steepest descent [208]). Therefore, we can write that

$$\begin{aligned}\phi(t) &= \ln \int d\Gamma \exp\Big\{-\sum_{j=1}^{n} F_j(t)\hat{P}_j(\Gamma)\Big\} \simeq \\ &\simeq \ln\Big[\exp\Big\{-\sum_{j=1}^{n} F_j(t)Q_j(t)\Big\} \int_{\mathcal{M}(t)} d\Gamma\Big] = \\ &= -\sum_{j=1}^{n} F_j(t)Q_j(t) + \ln \int_{\mathcal{M}(t)} d\Gamma\,,\end{aligned} \tag{4.145}$$

where the integration is over the manifold $\mathcal{M}(t)$ in phase space composed of the phase points $\Gamma \in \mathcal{M}(t)$ such that

$$\mathcal{M}(t) : Q_j(t) \leq \hat{P}_j(\Gamma) \leq Q_j(t) + \Delta Q_j(t)\,, \tag{4.146}$$

where ΔQ_j is of the order of $C_{jj}^{1/2}(t)$. Hence, using Eqs. (4.145) and (4.136) it follows that

$$\begin{aligned}S(t) &= k_B\phi(t) + \sum_{j=1}^{n} \mathcal{F}_j(t)Q_j(t) \simeq \\ &\quad - \sum_{j=1}^{n} \mathcal{F}_j(t)Q_j(t) + k_B \ln \int_{\mathcal{M}(t)} d\Gamma + \sum_{j=1}^{n} \mathcal{F}_j(t)Q_j(t)\,,\end{aligned} \tag{4.147}$$

and then

$$\boxed{S(t) = k_B \ln W\{Q_j(t)\}\,,} \tag{4.148}$$

where

$$W\{Q_j(t)\} = \text{extension of } \mathcal{M}(t) \equiv \text{ext}\{\mathcal{M}(t)\} \tag{4.149}$$

with extension meaning the measure of the hipervolume in phase space consisting of the phase points contained in $\mathcal{M}$. We recall that this is an approximate result, with an error of the order of the reciprocal of the square root of the number of degrees of freedom of the system, and therefore exact only in the thermodynamic limit.

Equation (4.148) represents the equivalent in IST of Boltzmann expression for the thermodynamic entropy in terms of the logarithm of the number of complexions compatible with the macroscopic constraints imposed on the system. It should be noticed that in IST they are given by the so-called informational set, the one composed of the constraints in the variational process in MaxEnt, that is the $\{Q_j(t)\}$, which are the average values of the set of mechanical variables $\{\hat{P}_j(t)\}$. Moreover, they are univocally related to the Lagrange multipliers (or set of intensive nonequilibrium thermodynamical variables) that also completely describe the macroscopic state of the system in IST, namely the set $\{k_B F_j(t) = \mathcal{F}_j(t)\}$.

The expression of Eq. (4.148) in the quantum level of description follows similarly, when we derive that

$$W\{Q_j(t)\} = \sum_{\tilde{n}\in\mathcal{M}(t)} 1 = \text{number of states in } \mathcal{M}(t)\ , \tag{4.150}$$

where $\tilde{n}$ is the set of quantum numbers which characterize the quantum mechanical state of the system, and $\mathcal{M}$ contains the set of states $|\tilde{n}\rangle$, such that

$$\mathcal{M}(t) : Q_j(t) \le \langle\tilde{n}|\hat{P}_j|\tilde{n}\rangle \le Q_j(t) + \Delta Q_j(t)\ , \tag{4.151}$$

where we have used the usual notations of bracs and kets and matrix elements between those states.

In terms of these results we can look again at the $\mathcal{H}$-theorem of section 4.8 and rewrite it in the form

$$\bar{S}(t) - \bar{S}(t_0) = k_B \ln \frac{\text{ext}\{\mathcal{M}(t)\}}{\text{ext}\{\mathcal{M}(t_0)\}} \ge 0\ , \tag{4.152}$$

where ext means the extension of the manifold $\mathcal{M}$ in the classical approach and number of states in the quantum approach fixed by the informational constraints. Evidently, Eq. (4.152) tells us that the extension of $\mathcal{M}$ increases in time, what can be interpreted in the following way:

Consider a closed system in an initially highly excited nonequilibrium state (to fix ideas let us think in terms of the photoinjected plasma in semiconductors), from which it evolves towards final equilibrium, an evolution governed by the kinetic equations [Eqs. (2.24). With elapsing time, as pointed out by Bogoliubov, subsets of correlations die down (in the case of photoinjected plasma implies the situation of processes of internal thermalization, nullification (decay) of fluxes, etc.) and a decreasing number of variables are necessary to describe the macroscopic state of the system. In IST this corresponds to a diminishing informational space, meaning of course a diminishing information, and, therefore, a situation less constrained with the consequent *increase of the extension of $\mathcal{M}$ and increase in informational entropy.*

Citing Jaynes, it is this property of the entropy — measuring our degree of information about the microstate, which is conveyed by data on the macroscopic thermodynamic variables — that made information theory such a powerful tool in showing us how to generalize Gibbs' equilibrium ensembles to nonequilibrium ones. The generalization could never have been found by those who thought that entropy was, like energy, a physical property of the microstate [202]. Also following Jaynes, $W(t)$ *measures the degree of control of the experimenter over the microstate,* when the only parameters the experimenter can manipulate are the usual macroscopic ones. At time t, when a measurement is performed, the state is characterized by the set $\{Q_j(t)\}$, and the corresponding phase volume is $W(t)$, containing all conceivable ways in which the final macrostate can be realized. But, since the experiment is to be reproducible, the region with volume $W(t)$ should contain at least the phase points originating in the region of volume $W(t_0)$, and then $W(t) \geq W(t_0)$. Because phase volume is conserved in the micro-dynamical evolution, it is a fundamental requirement on any reproducible process that the phase volume $W(t)$ compatible with the final state cannot be less than the phase volume $W(t_0)$ which describes our ability to reproduce the initial state [67].

4.13 Complementarity of Descriptions

The connection and interplay of the microscopic and macroscopic levels of description in matter, that is, between mechanics and thermodynamics, have been the object of discussion since the emergence of thermodynamics, as an offshoot of the Industrial Revolution, in last century. In particular, Leon Rosenfeld [204, 205] has argued that in this case is at work a kind of logical relationship to which the name of complementarity may be applied. This was conjectured by Niels Bohr [209], and is contained in an explicit form in Ilya Prigogine's work [125]. In Rosenfeld's words, it should characterize the mutual exclusiveness of the two descriptions: conditions allowing of a complete microscopic mechanical description of a system exclude the possibility of applying to the system any of the typical thermodynamic concepts; and, conversely, the macroscopic description in terms of the latter requires conditions of observation under which the mechanical parameters scape our control.

We consider here Rosenfeld-Prigogine's ideas in the framework of the Informational Statistical Thermodynamics so far described. It is applied to the study of a particular system consisting into an assembly of two subsystems of linear oscillators in mutual interaction (it constitutes an excellent model for the description of particular sets of collective elementary excitations in solids like, for example, polaritons, magnetoplasma waves, etc.).

Consider the system composed by two subsystems of harmonic oscillators, coupled through a particular interaction, as described by the Hamiltonian

$$\hat{H}(\hat{x},\hat{p};\hat{X},\hat{P}) = \hat{H}_{01}(\hat{x},\hat{p}) + \hat{H}_{02}(\hat{X},\hat{P}) + \hat{H}'(\hat{x},\hat{p};\hat{X},\hat{P}) \qquad (4.153)$$

where $\hat{x} \equiv \hat{x}_1,\ldots,\hat{x}_N$ and $\hat{X} \equiv \hat{X}_1,\ldots,\hat{X}_{N'}$ are the generalized coordinates of the two types of N and N' oscillators respectively and $\hat{p}$ and $\hat{P}$ are the corresponding sets of linear momenta. In this Eq. (4.153) $\hat{H}_{01}$ and $\hat{H}_{02}$ are the Hamiltonians of the free subsystems, namely

$$\hat{H}_{01} = \sum_{j=1}^{N} \frac{1}{2}(\hat{p}_j^2 + \omega_j^2 \hat{x}_j^2) \, ; \qquad (4.154)$$

$$\hat{H}_{02} = \sum_{\mu=1}^{N'} \frac{1}{2}(\hat{P}_\mu^2 + \Omega_\mu^2 \hat{X}_\mu^2) \, ; \qquad (4.155)$$

and $\hat{H}'$ is the interaction energy, which is taken in the form of the bilinear interaction,

$$\hat{H}' = \sum_{j,\mu} \Gamma_{j\mu} \hat{x}_j \hat{X}_\mu \, , \qquad j = 1,\ldots,N \, ; \; ;\mu = 1,\ldots,N' \, , \qquad (4.156)$$

where Γ stands for the coupling strength of the interaction.

Given the Hamiltonian of Eq. (4.153) we can proceed to build the informational mechanostatistical description of the system in MaxEnt-NESOM. We recall that the IST (or informational-statistical) entropy is given by [cf. Eq. (3.2)]

$$\bar{S}(t) = -\operatorname{Tr}\left\{\varrho(t) \ln \bar{\varrho}_\varepsilon(t,0)\right\} \equiv -\operatorname{Tr}\left\{\varrho_\varepsilon(t) \mathcal{P}_\varepsilon(t) \ln \varrho_\varepsilon(t)\right\} , \qquad (4.157)$$

expressed in terms of $\bar{\varrho}(t,0)$, the auxiliary coarse-grained distribution, and $\varrho_\varepsilon(t)$, the distribution that describes the macroscopic state of the system and its evolution in nonequilibrium conditions. They depend on the set of basic variables to be introduced for the description of the macroscopic state of the system, together with the set of accompanying Lagrange multipliers; let us call them, generically $\{Q_j(t)\}$ and $\{F_j(t)\}$ respectively. Moreover, $\mathcal{P}_\varepsilon(t)$ is the time-dependent projection operator, which projects on the so-called informational subspace composed by the dynamical variables used for the description of the system [cf. Fig. 2.1].

Two other relevant quantities for our purposes here are, one the already introduced informational-entropy production given by [cf. Eqs. (3.10) and (4.72)]

$$\bar{\sigma}(t) = \frac{d\bar{S}(t)}{dt} = \sum_{j=1}^{N} F_j(t)\frac{dQ_j(t)}{dt} , \tag{4.158}$$

and the others are the root-mean-square deviations $\Delta^2 Q_j(t)$ and $\Delta^2 F_j(t)$ [cf. Eqs. (4.134) and (4.140)]

Given $\phi(t)$ and $\bar{S}(t)$ of Eqs. (2.14) and (3.3), their differential coefficients give $Q_j(t)$ and $F_j(t)$, that is [cf. Eqs. (4.131) and (3.9)]

$$Q_j(t) = -\delta\phi(t)/\delta F_j(t) ; \qquad F_j(t) = \delta\bar{S}(t)/\delta Q_j(t) , \tag{4.159}$$

where δ stands for functional derivative [178]. Moreover, the second order functional derivatives allow to introduce the fluctuations of the basic variables, also providing for a generalization of Maxwell's relations to nonequilibrium situations, namely [cf. Eq. (4.135)]

$$\begin{aligned}\delta^2\phi(t)/\delta F_j(t)\delta F_k(t) \equiv C_{jk}(t) = -\delta Q_j(t)/\delta F_k(t) = \\ -\delta Q_k(t)/\delta F_j(t) = \int d\Gamma \Delta\hat{P}_j(t)\Delta\hat{P}_k(t)\bar{\varrho}(t,0) ,\end{aligned} \tag{4.160}$$

where

$$\Delta\hat{P}_j(t) = \hat{P}_j - \mathrm{Tr}\left\{\hat{P}_j\bar{\varrho}(t,0)\right\} = \hat{P}_j - Q_j(t) , \tag{4.161}$$

and $C_{jk}(t)$ is the matrix of correlations of the basic dynamical variables, which is symmetric, namely $C_{jk}(t) = C_{kj}(t)$, what, as noted, is a generalization to nonequilibrium situation (in the context of IST) of Maxwell's relations in Thermostatics. Furthermore, we can easily verified that

$$\delta^2\bar{S}(t)/\delta Q_j(t)\delta Q_k(t) = \delta F_j(t)/\delta Q_k(t) = \delta F_k(t)/\delta Q_j(t) , \tag{4.162}$$

and

$$\sum_l \frac{\delta^2\phi(t)}{\delta F_j(t)\delta F_l(t)} \frac{\delta^2\bar{S}(t)}{\delta Q_l(t)\delta Q_k(t)} = -\sum_l \frac{\delta Q_j(t)}{\delta F_l(t)} \frac{\delta F_l(t)}{\delta Q_k(t)} = -\delta_{jk}\,, \quad (4.163)$$

and then the second differential coefficients of the informational entropy are the elements of minus the inverse $\hat{C}^{(-1)}$ of the matrix of correlations $\hat{C}$. Let us next introduce the alternative definitions (thus introducing Boltzmann constant)

$$k_B\bar{S}(t) = S(t)\,; \qquad \mathcal{F}_j(t) = k_B F_j(t) = \delta S(t)/\delta Q_j(t)\,. \quad (4.164)$$

The fluctuation of the informational entropy is

$$\Delta^2 S(t) = \sum_{j,k} \frac{\delta S(t)}{\delta Q_j(t)} \frac{\delta S(t)}{\delta Q_k(t)} C_{jk}(t) = \sum_{j,k} C_{jk}(t)\mathcal{F}_j(t)\mathcal{F}_k(t)\,, \quad (4.165)$$

and that of the intensive variables $\mathcal{F}_j$ are

$$\Delta^2\mathcal{F}_j(t) = \sum_{k,l} \frac{\delta\mathcal{F}_j(t)}{\delta Q_k(t)} \frac{\delta\mathcal{F}_j(t)}{\delta Q_l(t)} C_{kl}(t) =$$
$$= k_B^2 \sum_{k,l} C_{jk}^{(-1)}(t) C_{jl}^{(-1)}(t) C_{kl}(t) = k_B^2 C_{jj}^{(-1)}(t)\,. \quad (4.166)$$

Moreover [cf. Eq. (4.141)]

$$\Delta^2 Q_j(t)\Delta^2\mathcal{F}_j(t) = k_B^2 G_{jj}(t)\,, \quad (4.167)$$

where [cf. Eq. (4.142)]

$$G_{jj}(t) = C_{jj}(t)C_{jj}^{(-1)}(t)\,. \quad (4.168)$$

In the particular case when the basic variables are uncorrelated, viz. $C_{jk} = 0$ for $j \neq k$, as in the case of equilibrium, then

$$\Delta^2 Q_j(t)\Delta^2\mathcal{F}_j(t) = k_B^2\,, \quad (4.169)$$

and

$$\left[\Delta^2 Q_j(t)\right]^{1/2}\left[\Delta^2\mathcal{F}_j(t)\right]^{1/2} = k_B\,. \quad (4.170)$$

Equations (4.167) or (4.170) have similarity with an uncertanty law as it is the case in Quantum Mechanics for the case of two noncommuting Hermitian operators.

Equation (4.167), after taken the square root, becomes

$$\boxed{\left[\Delta^2 Q_j(t)\right]^{1/2}\left[\Delta^2 \mathcal{F}_j(t)\right]^{1/2} = k_B\left[G_{jj}(t)\right]^{1/2}\,,} \tag{4.171}$$

where G_{jj} of Eq. (4.168) is a quantity equal to the product of the diagonal element j of the correlation matrix and that of its inverse, and equal to 1 in the case of uncorrelated variables, as shown. Clearly Eq. (4.171) resembles a kind of uncertainty principle in the way proposed by Rosenfeld [204, 205], which is *valid for arbitrarily far-from-equilibrium conditions and at any time during the evolution of the dissipative macrostate of the system.* We restate that the root-mean-square deviations $\Delta^2 \mathcal{F}_j(t)$ are to be understood in the same sense as is done in equilibrium, what is described in references [207, 210]; on the other hand the $\Delta^2 Q_j(t)$ represent the statistical fluctuations of the macrovariables, as already discussed in the previous section.

Let us return to the specific case of the system of oscillators characterized by the Hamiltonian of Eq. (4.153). We consider two different statistical descriptions of it: First, we consider the description in terms of

$$(I)\ \left\{\hat{H}_{01};\hat{H}_{02}\right\}\,;\ \left\{\beta_{1I}(t);\beta_{2I}(t)\right\}\,;\ \left\{E_1(t);E_2(t)\right\}\,, \tag{4.172}$$

for the dynamical variables, the Lagrange multipliers, and the macrovariables respectively, which consists of the collective variables corresponding to the energies of each of the subsystems, that is, we have here a kind of canonical description of each one in nonequilibrium conditions, with the auxiliary (coarse-grained) statistical probability distribution given then by

$$\bar{\varrho}_I(t,0) = \exp\left\{-\phi_I(t) - \beta_{1I}(t)\hat{H}_{01} - \beta_{2I}(t)\hat{H}_{02}\right\}\,. \tag{4.173}$$

Let us consider the equations of evolution for the two basic variables that is, the equations of the type of Eq. (2.24) for $Q_1(t) = E_1(t)$ and $Q_2(t) = E_2(t)$ in this case. As already noticed the right hand side of these equations is in fact a functional of the two Lagrange multipliers $\beta_{1I}(t)$ and $\beta_{2I}(t)$, on which $\bar{\varrho}(t,0)$ depends. But these Lagrange multipliers are related to the basic variables via the calculation of the average values, which in this case, and in a classical-mechanical approach, are, after a straightforward calculation, given by

$$E_1(t) = \mathrm{Tr}\left\{\hat{H}_{01}\bar{\varrho}(t,0)\right\} = N\beta_{1I}^{-1}(t)\,, \tag{4.174}$$

$$E_2(t) = \mathrm{Tr}\left\{\hat{H}_{02}\bar{\varrho}(t,0)\right\} = N'\beta_{2I}^{-1}(t)\,. \tag{4.175}$$

Therefore, the equations of evolution for the basic variables can be transformed into equations of evolution for the Lagrange parameters $\beta_{1I}(t)$ and $\beta_{2I}(t)$. This is done using the nonlinear transport theory that the method provides [64, 65, 100, 101, 113, 122, 187, 211], but resorting to the so-called second order approximation in relaxation theory [122], that is, the one that keeps the interaction up to second order (binary collisions), restricted then to weak interactions. Moreover, we take the limit of N' going to infinity, that is, the second system of oscillators plays the role of an ideal reservoir at, say, temperature T_0. The resulting system of equations of evolution, calculated in the Markovian limit of the NESOM-based kinetic theory [113, 122, 187, 212] are given by

$$\frac{d}{dt}E_1(t) = J_1^{(2)}(t)\,; \tag{4.176}$$

$$\frac{d}{dt}E_2(t) = J_2^{(2)}(t)\,, \tag{4.177}$$

where the collision operator $J^{(2)}$ is, we recall, given by

$$J_{1(2)}^{(2)}(t) = \int_{-\infty}^{0} dt' e^{\varepsilon t'} \int d\Gamma_1 d\Gamma_2 \left\{\hat{H}'(t')_0, \left\{\hat{H}', \hat{H}_{01(2)}\right\}\bar{\varrho}_I(t,0)\right\}\,, \tag{4.178}$$

with subindex nought in $\hat{H}'(t')_0$ indicating evolution in time under $\hat{H}_0$, and, we recall, ε is a positive infinitesimal that goes to zero after the calculation of the average has been performed; Γ_1 and Γ_2 are phase-space points of each subsystem. After some lengthy but straightforward algebra we find that

$$J_1^{(2)}(t) = \int_{-\infty}^{0} dt' e^{\varepsilon t'} \int d\Gamma_1 d\Gamma_2 \Big\{ \sum_{j,\mu,\nu} \Gamma_{j\mu}\Gamma_{j\nu} \Big[\hat{X}_\mu \hat{X}_\nu \cos(\omega_j t') \cos(\Omega_\mu t') + \hat{P}_\mu \hat{X}_\nu \Omega_\mu^{-1} \cos(\omega_j t') \sin(\Omega_\mu t') \Big] - \sum_{j,k,\nu} \Gamma_{j\mu}\Gamma_{j\nu}\Omega_\mu^{-1} [\hat{p}_j \hat{x}_k \cos(\omega_k t') \sin(\Omega_\mu t') + \hat{p}_j \hat{p}_k \omega_k^{-1} \sin(\omega_k t') \sin(\Omega_\mu t')] \Big\} \bar{\varrho}(t,0) \,. \quad (4.179)$$

A similar equation follows for $J_2^{(2)}(t)$ which we omit for the sake of brevity. The average values that appear in Eq. (4.179) are evaluated to obtain that

$$\langle \hat{p}_j \hat{x}_k | t \rangle = 0 \,; \qquad \langle \hat{p}_j \hat{p}_k | t \rangle = \delta_{jk} \beta_{1I}^{-1}(t) \,; \quad (4.180)$$

$$\langle \hat{P}_\mu \hat{X}_\nu | t \rangle = 0 \,; \qquad \langle \hat{X}_\mu \hat{X}_\nu | t \rangle = \delta_{\mu\nu} \Omega_\mu^{-2} \beta_{2I}^{-1}(t) \,; \quad (4.181)$$

and introducing these results into Eqs. (4.176) and (4.177) we find that

$$\frac{d}{dt} E_1(t) = \Gamma_1 \left[\beta_{1I}^{-1}(t) - \beta_{2I}^{-1}(t) \right] , \quad (4.182)$$

$$\frac{d}{dt} E_1(t) = -\frac{d}{dt} E_2(t) \,, \quad (4.183)$$

where

$$\Gamma_1 = \frac{\pi}{2} \sum_{j,\mu} \frac{\Gamma_{j\mu}}{\omega_j^2} \delta(\omega_j - \Omega_\mu) \,, \quad (4.184)$$

and Eq. (4.183) is a result of the conservation of energy in the global system. Moreover, taking into account Eqs. (4.174) and (4.175), we have a closed system of two equations for the two Lagrange parameters β_{1I} and β_{2I}.

After introducing the definitions

$$\beta_{1I}^{-1}(t) = k_B T_{1I}^*(t) \; and \; \beta_{2I}^{-1}(t) = k_B T_{2I}^*(t) \,, \quad (4.185)$$

with both T^* playing the role of nonequilibrium temperature-like variables (usually referred to as quasitemperatures [189]), using Eqs. (4.174),

(4.175), (4.182), and (4.183), we construct equations of evolution for the quasitemperatures, whose solution is

$$T^*_{1I}(t) = T_\infty + A_I e^{-t/\tau} \; ; \tag{4.186}$$

$$T^*_{2I}(t) = T_\infty = T_0 \; , \tag{4.187}$$

with the IST-entropy production being

$$\bar{\sigma}_I(t) = N\tau^{-1}\left[T^*_{1I}(t) - T^*_{2I}(t)\right]^2 / T^*_{1I}(t)T^*_{2I}(t) \; , \tag{4.188}$$

where T_∞ corresponds to the temperature when final thermal equilibrium is achieved, that is, at $t \to \infty$, when the temperature of system and reservoir coincide, and A_I is fixed by the initial conditions. In Eqs. (4.186) and (4.188) τ is a relaxation time given by N/Γ_1, i.e.

$$\frac{1}{\tau} \equiv \frac{1}{N}\sum_j \frac{1}{\tau_j} = \frac{1}{N}\sum_j \frac{\pi}{2}\sum_\mu (\Gamma^2_{j\mu}/\omega^2_j)\delta(\omega_j - \Omega_\mu) \; . \tag{4.189}$$

The entropy production of Eq. (4.188) is positive and becoming null when final equilibrium is achieved ($T^*_{1I} = T^*_{2I} = T_\infty = T_0$ for $t \to \infty$).

Consider now the description

$$(II)\;\left\{\hat{H}_{01}, \{\hat{x}_j\}, \{\hat{p}_j\}; \hat{H}_{02}\right\} ; \;\left\{\beta_{1II}(t), \{\varphi_{jII}(t)\}, \{\gamma_{jII}(t)\}; \beta_{2II}(t)\right\} ;$$
$$\left\{E_1(t), \{\bar{x}_j(t)\}, \{\bar{p}_j(t)\}; E_2(t)\right\} \tag{4.190}$$

which is a mixed one, involving the microscopic individual coordinates and momenta of the oscillators in subsystem 1 and the collective variables energy [as in (I); cf. Eq. (4.172)]. Therefore, the coarse-grained auxiliary distribution is in this case

$$\bar{\varrho}(t,0) = \exp\Big\{-\phi_{II}(t) - \beta_{1II}(t)\hat{H}_{01} - \beta_{2II}(t)\hat{H}_{02}$$
$$- \sum_{j=1}^N \left[\varphi_{jII}(t)\hat{x}_j + \gamma_{jII}(t)\hat{p}_j\right]\Big\}, \tag{4.191}$$

and, while $E_2(t)$ is again given in Eq. (4.175), we have now that

$$E_1(t) = N\beta^{-1}_{1II}(t) + \frac{1}{2}\sum_{j=1}^N \beta^2_{1II}(t)\left[\gamma^2_{jII}(t) + \omega^2_j\varphi_{jII}(t)\right] \; , \tag{4.192}$$

$$\omega_j^2 \bar{x}_j(t) = -\beta_{1II}^{-1}(t)\varphi_{jII}(t)\,, \tag{4.193}$$

$$\bar{p}_j(t) = -\beta_{1II}^{-1}(t)\gamma_{jII}(t)\,, \tag{4.194}$$

with

$$\bar{x}_j = \langle \hat{x}_j | t\rangle\,; \qquad \bar{p}_j = \langle \hat{p}_j | t\rangle\,. \tag{4.195}$$

We notice first that the average values of the basic variables can be written in the following forms

$$E_1(t) = \frac{N}{\beta_{1II}(t)} + \frac{1}{2}\sum_{j=1}^{N}\left[\bar{p}_j^2(t) + \omega_j^2 \bar{x}_j^2(t)\right] \tag{4.196}$$

and

$$E_2(t) = \frac{N'}{\beta_{2II}(t)}\,. \tag{4.197}$$

Proceeding as in the previous case (I), we derive and solve the equations of evolution to find that

$$T_{1II}^*(t) = T_\infty + A_{II}e^{-t/\tau}\,, \tag{4.198}$$

$$T_{2II}^*(t) = T_\infty = T_0\,, \tag{4.199}$$

$$\bar{x}_j(t) = (c_j/\omega_j)\exp(-t/2\tau_j)\cos(\omega_j t + \theta_j)\,, \tag{4.200}$$

$$\bar{p}_j(t) = -\bar{x}_j(t)/2\tau_j + c_j\exp(-t/2\tau_j)\sin(\omega_j t + \theta_j)\,, \tag{4.201}$$

$$\bar{\sigma}(t) = N\tau^{-1}\frac{[T_{1II}^*(t) + T_{2II}^*(t)]^2}{T_{1II}^*(t)T_{2II}^*(t)} + \frac{f(t)}{k_B T_{2II}^*(t)}\,, \tag{4.202}$$

where c_j and θ_j are determined by the initial conditions,

$$f(t) = \sum_{j=1}^{N}\left[(\omega_j^2/\tau_j)\bar{x}_j^2(t) + \Lambda_j\bar{x}_j(t)\bar{p}_j(t)\right]\,, \tag{4.203}$$

τ_j is defined in Eq. (4.189), and

$$\Lambda_j = \sum_{\mu=1}^{N'}\Gamma_{j\mu}^2(\Omega_\mu^2 - \omega_j^2)^{-1}\,, \tag{4.204}$$

and we recall that the subsystem 2 acts as an ideal reservoir at constant temperature $T^*_{2II}(0) = T_\infty = T_0$.

Next we compare both descriptions (I) and (II) [cf. Eqs. (4.173) and (4.191)] using, evidently, the same initial conditions in both cases. We fix the initial energies of both systems, which using Eqs. (4.174), (4.175), (4.192) and (4.194) can be written in the form

$$(I)\ \Delta E = E_1(0) - Nk_BT_\infty = \delta q_{1I}\,, \tag{4.205}$$

$$\delta q_{1I} = Nk_B\left[T^*_{1I}(0) - T_\infty\right] = Nk_BA_I\,; \tag{4.206}$$

$$(II)\ \Delta E = E_1(0) - Nk_BT_\infty = \delta q_{1II} + \delta w_{1II}\,, \tag{4.207}$$

$$\delta w_{1II} = \sum_{j=1}^{N}(c_j^2/2)\,, \tag{4.208}$$

$$\delta q_{1II} = Nk_B\left[T^*_{1II}(0) - T_\infty\right] = Nk_BA_{II}\,. \tag{4.209}$$

We recall that the energy of the reservoir E_2 is fixed by its temperature, $T_0 = T_\infty$, and for the sake of simplicity without losing generality, we have chosen the initial conditions $\bar{x}_j(0) = 0$ and $\bar{p}_j(0) = c_j$. These Eqs. (4.205) to (4.209) provide the initial energy in excess of the values in final equilibrium, ΔE, which is composed of two terms, one δq which we call a "heat-like contribution" and the other δw dubbed a "work-like contribution".

We proceed to compare both descriptions for which purpose, first, we resort to a quantum description, more appropriate for a full analysis in what follows.

In a quantal approch the two descriptions we have introduced are, in terms of the dynamical quantities, in the first case [cf. Eq. (4.172)] composed of the Hamiltonian operators

$$\hat{H}_{10} = \sum_{j=1}^{N}\hbar\omega_j(a_j^\dagger a_j + \frac{1}{2})\,, \tag{4.210}$$

$$\hat{H}_{20} = \sum_{\mu=1}^{N'}\hbar\Omega_\mu(b_\mu^\dagger b_\mu + \frac{1}{2})\,, \tag{4.211}$$

where $a_j^\dagger(a_j)$, $b_\mu^\dagger(b_\mu)$ are annihilation (creation) operators in the corresponding states. In the second case [cf. Eq. (4.190)], besides the two Hamiltonians above, are incorporated the quantities a_j and $a_j^\dagger$, which, through appropriate linear combinations produce the operators for coordinate and momentum of the oscillator in the second quantization representation. The auxiliarly coarse-grained statistical operators are in this case

$$\bar{\varrho}_I(t,0) = \exp\left\{-\phi_I(t) - \beta_{1I}(t)\hat{H}_{10} - \beta_{2I}(t)\hat{H}_{20}\right\}, \tag{4.212}$$

$$\bar{\varrho}_{II}(t,0) = \exp\Big\{-\phi_{II}(t) - \beta_{1II}(t)\hat{H}_{10} - \beta_{2II}(t)\hat{H}_{20} - \big(\sum_{j=1}^{N} f_j(t)a_j + \text{H.c.}\big)\Big\}, \tag{4.213}$$

where ϕ, β, f_j are the corresponding Lagrange multipliers. The macrovariables are

$$E_1(t) = \text{Tr}\left\{\hat{H}_{10}\bar{\varrho}_I(t,0)\right\}; \qquad E_2(t) = \text{Tr}\left\{\hat{H}_{20}\bar{\varrho}_I(t,0)\right\} \tag{4.214}$$

in the first description and in the second are

$$E_1(t) = \text{Tr}\left\{\hat{H}_{10}\bar{\varrho}_{II}(t,0)\right\}; \qquad E_2(t) = \text{Tr}\left\{\hat{H}_{20}\bar{\varrho}_{II}(t,0)\right\}, \tag{4.215}$$

together with

$$\langle a_j|t\rangle = \text{Tr}\left\{a_j\bar{\varrho}_{II}(t,0)\right\}; \; \langle a_j|t\rangle^* = \text{Tr}\left\{a_j^\dagger\bar{\varrho}_{II}(t,0)\right\}. \tag{4.216}$$

It can be noticed that the statistical operator $\bar{\varrho}_{II}(t,0)$ of Eq. (4.213) can be expressed in terms of only the population operators for a new set of quantities, say $\tilde{a}$ once the Glauber-like transformation

$$a_j = \tilde{a}_j + \langle a_j|t\rangle, \tag{4.217}$$

is performed. The calculations are then greatly simplified, and it can be shown that

$$\nu_j = \text{Tr}\left\{a_j^\dagger a_j\bar{\varrho}_{II}(t,0)\right\} = \left\{\exp\left[\beta_{1II}(t)\hbar\omega_j\right] - 1\right\}^{-1} + |\langle a_j|t\rangle|^2, \tag{4.218}$$

$$\langle a_j|t\rangle = -f^*(t)/\beta_{1II}(t)\hbar\omega_j. \tag{4.219}$$

Using the results listed above, after some algebra, the different statistical-thermodynamic functions can be calculated to obtain in the first description that

$$E_1(t)/N = \frac{1}{2}\hbar\omega_0 \coth(\frac{1}{2}\beta_{1I}(t)\hbar\omega_0) , \tag{4.220}$$

$$\phi_I(t)/N = -\ln\left[2\sinh(\frac{1}{2}\beta_{1I}(t)\hbar\omega_0)\right] , \tag{4.221}$$

$$\begin{aligned} \bar{S}_I(t)/N &= [\phi_I(t)/N] + \beta_{1I}(t)\,[E_1(t)/N] = \\ &= -\ln\left[2\sinh(\frac{1}{2}\beta_{1I}(t)\hbar\omega_0)\right] + \frac{1}{2}\hbar\omega_0\coth(\frac{1}{2}\beta_{1I}(t)\hbar\omega_0) . \end{aligned} \tag{4.222}$$

In the derivation of these equations we have taken a unique frequency for all the oscillators, and the second system is taken as an ideal reservoir. In the second description we find that

$$E_1(t)/N = \frac{1}{2}\hbar\omega_0 \coth(\frac{1}{2}\beta_{1II}(t)\hbar\omega_0) + \beta_{1II}^{-2}(t)(\Lambda/k_B^2) , \tag{4.223}$$

$$\phi_{II}(t)/N = -\ln\left[2\sinh(\frac{1}{2}\beta_{1II}(t)\hbar\omega_0)\right] - \beta_{1II}^{-1}(t)(\Lambda/k_B^2) , \tag{4.224}$$

$$\begin{aligned} \bar{S}_{II}(t)/N &= [\phi_{II}(t)/N] + \beta_{1II}(t)\,[E_1(t)/N] = \\ &= -\ln\left[2\sinh(\frac{1}{2}\beta_{1II}(t)\hbar\omega_0)\right] + \frac{1}{2}\hbar\omega_0\coth(\frac{1}{2}\beta_{1II}(t)\hbar\omega_0) , \end{aligned} \tag{4.225}$$

where $\Lambda = \Delta w/N$ and

$$\Delta w = k_B^2 \sum_{j=1}^{N} \left|f_j\right|^2 . \tag{4.226}$$

We use these results in the numerical calculations, proceeding in the same way as done in the classical approach.

We define what we call an *order parameter* given by

$$\Delta(t) = \left[\bar{S}_I(t) - \bar{S}_{II}(t)\right]/\bar{S}_I(t) = K(t)/\bar{S}_I(t) , \tag{4.227}$$

where we have introduced

$$K(t) = -\operatorname{Tr}\left\{\bar{\varrho}_I(t,0)\ln\bar{\varrho}_I(t,0) - \bar{\varrho}_{II}(t,0)\ln\bar{\varrho}_{II}(t,0)\right\}, \tag{4.228}$$

namely, an analog of Kullback's information measure [213–215], which is interpreted as a measure of the gain in information in the description using $\bar{\varrho}_{II}$ in comparison with the one using $\bar{\varrho}_I$. In Fig. 4.3 it is shown the evolution of Δ for the choice $T_0 = 300K$; $\Delta E = 0.1Nk_BT_0$; $\delta w_{1II} = 0.1\Delta E$; and all $\hbar\omega_j$ equal to $35meV$. The IST entropy in description II is smaller than in description I, as expected, since the former carries more information, but they asymptotically coincide once the final temperature in equilibrium is achieved, as it should. Further considerations on the informational entropy and its production are given in [166, 176, 216, 217], and a geometrical-topological discussion of the method is due to Balian et al. [181].

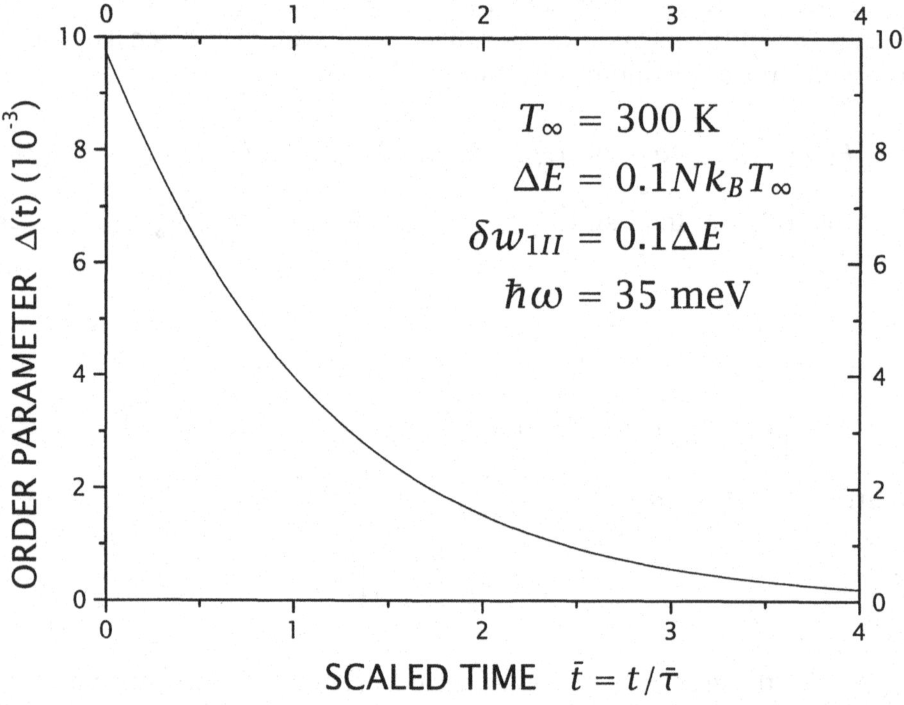

Figure 4.3: Evolution in time of the order parameter of Eq. (4.227).

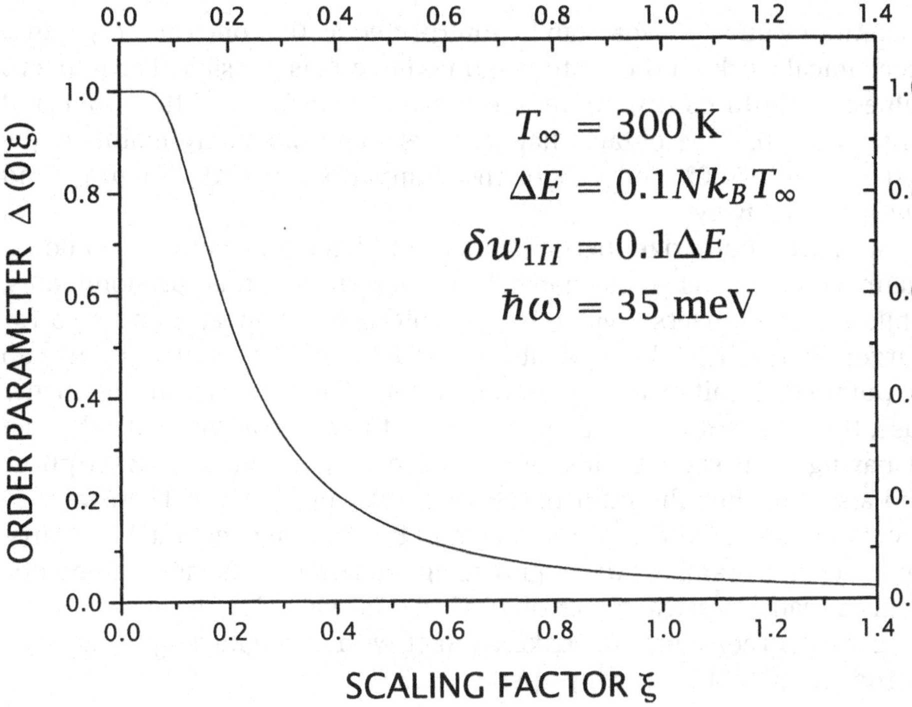

Figure 4.4: Dependence of the order parameter of Eq. (4.227) for $t = 0$ on a scaled Boltzmann constant.

Taking into account all these results together with the relationship of Eq. (4.171), we explore the role of Boltzmann constant resorting to introducing at a given time (say the inital one $t = 0$) in the expression for Δ of Eq. (4.227) a scaling ξ of the Boltzmann constant (writting ξk_B), with ξ varying from zero to infinity. The resulting ξ-dependent $\Delta(0|\xi)$ is shown in Fig. 4.4. It is verified that $0 \leq \Delta \leq 1$, with Δ going to one for ξ going to zero and Δ going to zero for ξ going to infinity, implying in maximum information gain and no information gain at all respectively. For the numbers used (and we recall that δw_{1II} is 10% of the input of exciting energy ΔE, while δq_{1II} is 90% of ΔE), for $\xi = 1$, that is the real case in nature for $k_B = 8.617 \times 10^{-5}$ eV/K, the information gain is roughly 1% of $\bar{S}_I$. Moreover, it follows that for ξ small the "heat-like contribution" δq_{1II} goes to zero, while, the "work-like contribution" δw_{1II} acquires the maximum value ΔE, what can be interpreted as that one can only pump mechanical work on the system and no heating is possible. For nonzero value of ξ both contributions are present, and for ξ of the order and larger than 0.5, we obtain that they very approximately maintain the distribution of 90% and 10% of the pumped energy ΔE, for δq_{1II} and δw_{1II} respectively.

We may summarize these results as implying that for a "small Boltzmann constant" is a mechanical-like description that predominates, while for the universal value of the Boltzmann constant, and also for "larger values of it", both "heat-like" and "work-like" contributions can be pumped simultaneaously on the system. Furthermore, in the former case, the informational entropy $\bar{S}_{II}$ tends to zero in accord with the fact of having what can be considered as a purely mechanical description, and also vanishes the informational-entropy production: Then we may say that in such limiting situation ("null Boltzmann constant") no statistical thermodynamics exists, quite in agreement with Jaynes'statements, in his already classical paper of 1965 [67] (see also [202]).

On the other hand, to make contact with Prigogine's approach, we introduce in IST the entropy operator [137]

$$\hat{S}(t) = k_B\hat{\hat{S}} = -k_B\mathcal{P}_\varepsilon(t)\ln\varrho(t) = -F_0(t) - \sum_{j=1}^{n} F_j(t)\hat{P}_j\,, \qquad (4.229)$$

where $F_0(t) = k_B\phi(t)$ and $\mathcal{P}_\varepsilon(t)$ is the time-dependent projection operator present in Eq. (4.157) and defined in Eq. (2.18) (which projects at each time t over the subspace defined by the basic set of dynamical variables, the so-called informational subspace; see also [181]). The statistical average with $\varrho_\varepsilon(t)$ of this entropy operator is the IST entropy of Eq. (4.157). Furthermore, if we indicate by $\mathcal{L}$ the Liouville operator of the system, then it follows that

$$k_B\hat{\hat{\sigma}}(t) \equiv \mathrm{i}\mathcal{L}\hat{S}(t) = \sum_{j=1}^{n} F_j(t)\mathrm{i}\mathcal{L}\hat{P}_j\,, \qquad (4.230)$$

which introduces the entropy production operator, $\hat{\hat{\sigma}}$, whose average is the IST entropy production of Eq. (4.158). The connection of the entropy operator of Eq. (4.229) and the one introduced in a general form by Prigogine is made through the identification $-k_B\mathcal{P}_\varepsilon(t)\ln\varrho_\varepsilon(t) \equiv \hat{M}$, with operator $\hat{M}$ defined in [125]. Let us take the commutator of the Liouville operator and the entropy operator, and next the average value of it, to obtain that

$$\mathrm{Tr}\left\{[\mathrm{i}\mathcal{L},\hat{S}(t)]\varrho_\varepsilon(t)\right\} = \frac{1}{\mathrm{i}\hbar}\mathrm{Tr}\left\{[\hat{S}(t)\varrho_\varepsilon(t),\hat{H}] - \hat{S}(t)[\varrho_\varepsilon(t),\hat{H}]\right\} = \\ = \mathrm{Tr}\left\{\mathrm{i}\mathcal{L}\hat{S}(t)\varrho_\varepsilon(t)\right\} = k_B\bar{\sigma}(t)\,. \quad (4.231)$$

According to Prigogine [125], the non-null commutator in this Eq. (4.231) leads to a complementarity principle that implies that either we consider eigenfunctions of the Liouville operator to determine the mechanical evolution of the system or we consider eigenfunctions of the entropy operator [137], but they do not have common eigenfunctions.

On the basis of the results presented above we can say that they point out to the plausibility that the incommensurability of the Liouville operator (mechanical level) and the entropy operator (thermodynamical level) implies in a kind of uncertainty relation, or, more appropriately, a kind of measure of incompleteness of descriptions:

> *A simultaneous determination of the informational content of the solutions of the equations of evolution of the macrostate and a detailed microscopic positioning (point in phase space or quantum state) is not possible.*

This fact is governed by the presence of Boltzmann constant, as quantified in Eq. (4.171) and Eq. (4.231) (see Fig. 4.4, where the role of k_B — scaled by the factor ξ — to characterize this complementarity principle is evidenced).

It has been argued [218] that k_B introduces the influence on the microscopic level of the experiment at the macroscopic level. Heat and work are considered as intrinsic properties of matter, and heat flux as a movement of "thermal charges" under the action of a gradient of temperature. In this context k_B may then be — as reinforced by the results in this section — considered as a "quantum of thermal charge", namely, the minor amount of heat to be displaced by unit of temperature gradient. It would *represent the unit of measure of the uncertainty of description of the mechanical state on the basis of the given reduced macroscopic characterization of the system.* This point has also been stressed by L. Tisza [21].

Hence, it may be argued that, as Planck's constant defines the interaction between the quantum system and the measurement device as nondecomposable, Boltzmann constant defines the microscopic and macroscopic descriptions also as nondecomposable [127]. In this case, we reiterate, it is at work a kind of logical relationship to which the name of *complementarity* — as an extension of Bohr's ideas — may be applied [204, 205]: As shown (and we stress that this is in the realm of IST) it can be characterized by a kind of uncertainty relation [cf. Eq. (4.171)] and the interplay of two noncommutative operators [cf. Eq. (4.231)] (one associated to the mechanical level of description and the other to the thermal one), with Boltzmann constant playing the role of an elementary quantum of heat transfer and responsible for the necessary duality of descriptions.

4.14 Summary of Chapter 4 and further Considerations

Summarizing, in this lengthy Chapter 4 we have presented first an idea concerning the construction of the so-called Informational Statistical Thermodynamics, but concentrating the attention in a particular nonequilibrium-thermodynamic state-like function dubbed the informational-statistical entropy or entropy in IST, and its properties. In short, after the definition of this informational entropy within the conceptual point of view and framework of the MaxEnt-NESOM, we analized:

(a) The nonequilibrium equations of state arising out of the differential coefficients of the informational entropy. This establishes an important connection between the informational basic macrovariables and the Lagrange multipliers that the variational method introduces, elevating the latter to the level of nonequilibrium-thermodynamic intensive variables.

(b) A generalized $\mathcal{H}$-like theorem, in Jancel's sense, is derived showing that the informational entropy increases in time from its initial value at the time of initial preparation of the system in the experiment to be considered. This leads to what we have called a weak principle of informational-entropy production, i.e. a time-integrated one, since so far there is no demonstration probing it to be non-negative at any given time. As noted in section 4.8 we think to be possible for the informational-entropy production to be in fact always (at any time during the irreversible evolution of the system) non-negative whereas Zubarev-Peletminskii selection rule is exactly satified (i.e. one has a complete set of macroscopic (or mesoscopic) variables for the description of the nonequilibrium thermodynamic state of the system), but not being satisfied when a truncation, and thus an approximation, is introduced (so information is increasing at certain times because of the additional constraints that are introduced to enforce the truncation).

(c) On the basis of considerations on the convexity of the informational entropy in the space of macrovariables $\{Q_j\}$ (ensured by the process of maximization), it is possible to analize the informational entropy production to derive that, (i) there follows a particular criterion for evolution; and in the case of steady states: (ii) a criterion for their (in)stability, and (iii) a theorem of minimum informational-entropy production in steady states near equilibrium (linear regime) are verified.

This item (iii) excludes the possibility of existence of complex behavior (synergetic self-organization in dissipative structures) in the Onsagerian regime. In the nonlinear domain (outside Onsager's regime) the other two criteria [(i) and (ii)] enter into action and instability against the formation of Prigogine-like dissipative structures is possible (when the matrix of kinetic coefficients contains asymmetric contributions implying in that Onsager symmetry relations are no longer satisfied).

(d) A generalized Clausius-like relation may be introduced in IST, through an appropriate generalization of a heat function and its change in irreversible processes, and the use of the concept of quasitemperature (per subsystem). This establishes a certain relationship with Carnot's approach. All these results go over those of linear and nonlinear phenomenological classical thermodynamics, and elemental kinetic theories (Boltzmann equation, classical hydrodynamics, etc.) in appropriate asymptotic limits.

(e) A theory of fluctuations is derived and the generalization to nonequilibrium conditions of Maxwell relations (relationships between the differential coefficients of the informational entropy) are obtained. It also follows a kind of uncertainty principle between the fluctuations of the macrovariables and the root-mean-squared deviations of the Lagrange multipliers. This leads to a proposal of something akin to a complementarity principle between macro/micro descriptions, in the way advanced by Rosenfeld and Prigogine, and also Krylov, in the sense that it is impossible to specify minutely the more precise location of a system within the phase space region (in the classical case) delimited by the macroscopic characteristics of the system.

(f) Also on the basis of the results of item (d), it is possible to derive, in the thermodynamic limit, an expression for the informational entropy formally identical to the famous Boltzmann's expression, namely, the informational entropy is proportional to the logarithm of the measure of the volume of the region in phase space consisting of the phase points representing the microstates compatible with the given average values of the mechanical quantities that define the informational space (that is the values of the macrovariables in Gibbs' space to within an indeterminacy given by their fluctuations). In the quantum case the extension of volume in phase space is to be replaced by the number of quantum states compatible with the informational constraints. Such volume increases its extension in time (or the number of quantum states increases in time) along the irreversible evolution of the system as a result of a loss of information as times elapses and the system tends to final equilibrium when it attains a maximum value.

Chapter 5

Concluding Remarks

In this book we have briefly and partially considered aspects of the present status of phenomenological irreversible thermodynamics and nonequilibrium statistical mechanics. More precisely we have described phenomenological theories of irreversible thermodynamics and the possibility to place them under the aegis of the Nonequilibrium Statistical Operator Method in the form of the so-called Informational Statistical Thermodynamics.

After general comments in the Introduction (stressing the relevance of the subject on what can be termed as nonlinear macrophysics, and its connection with complexity), in Chapter 1 we have discussed several aspects — the foundations, limitations and prospects — of phenomenological irreversible thermodynamics. In Chapter 2 we have briefly shown how to perform the construction of a nonequilibrium ensemble formalism in the form of MaxEnt-NESOM, and its connection with Jaynes' Predictive Statistical Mechanics. The variational procedure MaxEnt is introduced in such a way to include memory effects and an *ad hoc* hypothesis to ensure irreversible behavior in the evolution of the macroscopic state of the system from an initial condition of preparation. As noted, Predictive Statistical Mechanics is founded on Bayesian methods. According to Jaynes, the question of what are the theoretically valid, and pragmatically useful, ways of applying probability theory in science has been approached by Sir Harold Jeffreys [219], in the sense that he stated the general philosophy of what scientific inference is and proceeded to carry both the mathematical theory and its practical implementations [104, 105]. Also, Jaynes [11, 13, 103] and P. W. Anderson [220] maintain that what seems to be the most appropriate probability theory for the sciences is the Bayesian approach. The Bayesian interpretation is that probability is the degree of belief which is consistent to hold in a proposition being true, according to which other conditioning propositions are taken as true [221]. Or, according to Anderson [220], "What Bayesian does is to focus one's attention on the question one wants to ask of the data: It says in effect, How do these data affect my previous knowledge of the situation? It is sometimes called maximum likelihood thinking, but the essence of it is to clearly identify the possible answers, assign reasonable a priori probabilities to them and then ask which answers have been done more likely by the data". As we have seen there is a constructive criterion for deriving the probability assignment for the problem on the basis of the available information, namely MaxEnt. The fact that a certain probability distribution maximizes the Gibbs entropy [cf. Eq. (2.5)] subject to certain constraints representing our incomplete information, is the fundamental property which justifies the use of that distribution for inference; it agrees with everything that is known, but carefully avoids assuming anything that is not known. In that way it enforces — or gives a logico-mathematical viewpoint — to the principle of economy in logic, known as Occam's razor, namely, "Entities are not to be multiplied except of necessity." The resulting Jaynes' Predictive Statistical Mechanics provides laws of thermodynamics, not on the basis of the usual viewpoint of mechanical trajectories and ergodicity of classical deductive reasoning, but by the goal of using inference from incomplete information rather than deduction: the MaxEnt distribution represents the best predictions we are able to make from the information we have [11, 13, 71].

As noticed along the previous Chapters, MaxEnt-NESOM, on its own, seems to have an important role in the generation of statistical-mechanical foundations for phenomenological irreversible thermodynamics. To begin with, this applies already to Classical (or Onsagerian) Irreversible Thermodynamics: one can rederive the basic laws regarding evolution and (in)stability of dissipative many-body systems, and to retrieve results as linear response function theory and existing transport theories. It has been shown in Chapter 3 that it also covers the case of Extended Irreversible Thermodynamics. This is possible as a result of: (1) the application of Bogoliubov's concept of a hierarchy of relaxation times, with the associated contraction of the description, which is fundamental for the construction of the MaxEnt-NESOM, together with, (2) the separation in the system Hamiltonian [Cf. Eq. (2.3)] of a secular part $\hat{H}_0$ plus the weak interactions in $\hat{H}'$, responsible for the dissipative collisional processes, and (3) the iterative application of the Zubarev-Peletminskii closure condition [Cf. Eq. (2.4)]. A MaxEnt-NESOM entropy function, or informational entropy, for the system, depending on a coarse-grained probability distribution function suitably expressed in term of the basic variables, is introduced. The main properties of this MaxEnt-NESOM entropy is that it has associated a linear differential form in the thermodynamic Gibbs space [Cf. Eq. (3.4)] (thus providing a generalized Gibbs relation), and that its time derivative, the MaxEnt-NESOM entropy-production function (rate of change in time of the informational entropy) [Cf. Eq. (3.10)], may be regarded as the natural extension of the entropy production in Classical Irreversible Thermodynamics, which is semipositive definite in the linear regime around equilibrium. It should be stressed that at present it has not been found a way to show such character for the general case of systems arbitrarily away from equilibrium, on the basis of its expression as given by Eq. (3.11), except for the weak condition of Eqs. (3.12) and (3.13).

The theory is completed with an accompanying nonlinear quantum kinetic theory that provides the equations of evolution for the basic variables (cf. Eqs. (2.23) and (2.24) and also Eqs. (4.31) in the case of the nonequilibrium generalized grand-canonical ensemble), which involves the presence of a collision operator quite difficult to handle mathematically [222], but that can be rearranged in a more practical way suitable to introduce appropriate approximations [64, 65, 122]. These transport equations are substantial generalizations of the Maxwell-Cattaneo-Vernotte equations of Extended Irreversible Thermodynamics. In the theory, all relaxation times and transport coefficients can be, in principle, calculated from the microscopic dynamics of the system averaged in terms of the coarse-grained distribution $\bar{\varrho}$ of Eq. (2.13) [see also Eq. (4.77)]. In its full version the kinetic equations are highly nonlinear, nonlocal (i.e. contain space correlations), history dependent (i.e. have memory but of a fading character), and containing all the fluxes involved up to any order. We reinforce the fact that, going beyond Classical (Onsagerian) Irreversible Thermodynamics, the fluxes have been promoted to basic variables, and so the theory provides rigorous equations of evolution for them, in place of the auxiliary (phenomenological) equations as Eqs. (1.10). In a low order approximation (local in space and time and linear in the fluxes) equations of the Maxwell-Cattaneo-Vernotte-type, like Eq. (1.12), are obtained. This kinetic theory arising out of the MaxEnt-NESOM has a quite large range of applicability and can be considered a far-reaching generalization of earlier theories, like Boltzmann's transport theory [97], Mori's generalized Langevin equations [77], etc.

In section 4.8 has been presented the derivation of a criterion for irreversibility, consisting in a generalized $\mathcal{H}$-theorem. According to it the informational entropy is always larger, or at most equal, to Gibbs entropy, as a result of the coarse-graining imposed by the use of the contracted macroscopic description of the system [cf. Eq. (3.2) and Fig. 2.1]. More generally it appears to be possible to show that there exists a hierarchy of values of entropy which increase in correspondence with a decrease in the number of elements of the basic set of variables, namely, the different degrees of constraints. In that way it seems to be possible to offer a possible characterization of Rosenfeld's complementarity principle between (microscopic) mechanical and (macroscopic) statistical descriptions [206]. Results reported by Llebot and Criado-Sancho [196] already show that, even though the entropy is an increasing function in time, the entropy production may be negative along short intervals of time, but with a tendency to monotonically increase in time as the number of basic variables is increased. Nevertheless, we stress that at present there is not a conclusive demonstration that the MaxEnt-NESOM entropy production is instantaneously positive definite. The proof we presented only ensures what we have called a weak principle of positive informational entropy production, which states that as the system evolves in time the local informational-entropy production is predominantly positive definite.

Other relevant properties of the informational entropy has been described in section 4.9. On the one side the derivation of a principle of evolution which can be considered a generalization in IST of the one derived by Glansdorff and Prigogine in nonlinear classical thermodynamics. It consists in that along the trajectory of evolution of the macroscopic state of the system the part of the rate of variation in time of the informational entropy production, arising out of the change in time of the Lagrange multipliers, is alway non-positive (in the case of classical thermodynamics corresponds to the change in the thermodynamic forces). Another result is a criterion for (in)stability of steady-states, based on the sign of the so-called excess of internal entropy production function — that is, the difference between the internal entropy production near a steady state and its value at the given steady state: this is also a generalization in IST of Glansdorff-Prigogine criterion in nonlinear classical thermodynamics. Moreover, in the particular limiting case of steady states near equilibrium (the Onsager's domain or strictly linear regime) there follows a theorem of minimum production of informational entropy, once again a generalization in IST of the one provided by Prigogine in linear (Onsager) classical thermodynamics.

In section 4.10 we have derived a kind of generalized Clausius relation, meaning to express the increase in time of the informational entropy, along the trajectory of evolution of the thermodynamic state of the system, in terms of an integration of an infinitesimal (nonexact differential) of a pseudo-heat function divided by the space and time dependent quasitemperature. This result may, in a somewhat unconventional way — or may be in a whimsical way - be considered as the reversible exchange of heat, in a series of infinitesimal quasi-static processes, with a pseudo-reservoir having a local and instantaneuos temperature equal to the quasitemperature of the system. We recall that in the limit when IST recovers as a particular case classical Onsagerian irreversible thermodynamics, the quasitemperature goes over the local equilibrium temperature.

The section 4.11 has been devoted to the derivation of the covariances (or direct and cross-fluctuations) of the basic variables. A generalized matrix of correlations is obtained, whose elements can also be expressed as the second differentials of the functional ϕ of Eq. (2.14) [which, as noticed, plays the role of the logarithm of a nonequilibrium partition function, say $\phi(t) = \ln \bar{Z}(t)$] with respect to the Lagrange multipliers. The inverse of this matrix of correlations has as elements the second order variational derivatives of the informational entropy with respect to the basic macro-variables. These matrices are symmetric, what is a manifestation of Maxwell relations in nonequilibrium conditions. These results appear to provide a framework for a detailed study of fluctuations in dissipative steady states, and the connection with the attractive question of synergetic selforganization in far-from-equilibrium open systems.

On the basis of the results of the section 4.11, in section 4.12 we have shown that in the thermodynamic limit the informational entropy (in units of Boltzmann constant) acquires at each time, along the irreversible evolution of the system, an expression akin to the famous Boltzmann expression, that is, can be written as Boltzmann constant times the logarithm of the extension (in units of Planck constant to the power of the number of degrees of freedom) of the volume in phase space of the region of phase points describing the microstates compatible with the constraints imposed by the given values of the informational variables (the average value of the basic dynamical variables or micromechanical quantities), to within an indetermination given by their root-mean-squared deviation. The validity of the expression extends to the quantum level of description when the extension in phase space is substituted by the number of quantum mechanical states limited by the constraints. According to the $\mathcal{H}$-theorem of section 4.7, it increases in time, meaning that the number of states compatible with the informational constraints increases in time, a result of the loss of information that occurs along the irreversible evolution of the system.

After this summary of the content of the main text we close this concluding remarks with some general comments on the questions considered, namely, the nonequilibrium ensemble method and an entropy-like state function for dissipative systems.

As noted by L. Sklar [1], some attempts to generalize nonequilibrium statistical mechanics, beyond its origins in Maxwell's and Boltzmann's work, are available. However, there is no single, coherent, fully systematic theory that all will agree to that constitutes the correct core of nonequilibrium theory. Also, as Oliver Penrose [88] pointed out, even the well accepted and respectable equilibrium theory has its conceptual difficulties. We have resorted to the Non-Equilibrium Statistical Operator Method, a generalization to arbitrary nonequilibrium conditions of Gibbs' ensemble ideas, which, as cited in the Introduction, has an appealing structure and is most effective for dealing with nonlinear transport processes. Such statistical approach has had precursors, in one way or other as noticed in the Introduction, in the work of Kirkwood, Green, Zwanzig, Mori, and others. Based on fundamental ideas put forward by Nicolai N. Bogoliubov, it was largely developed by the Russian School of statistical mechanics, mainly by Dimitri N. Zubarev, S. V. Peletminskii and others. The original proposal by Zubarev, based on heuristic considerations, was placed by Zubarev and Kalashnikov within the scope of the variational method in statistical mechanics pioneered by Jaynes, and related to Information Theory initiated by Shannon with his mathematical theory of communications. For that reason we consider the method as encompassed within the scope of Jaynes' proposal of a Predictive Statistical Mechanics. It is based on Bayesian methods in probability theory and the principle of maximization of the informational entropy defined in the sense provided by Shannon, already referred to as MaxEnt for short. It was pointed out by Jaynes that maybe we ought to begin to understand that *science is really information organized in a particular way.*

Needless to say that quite difficult conceptual questions are open, for example how to determine up to what point the information reside in us (the limitation of the observer), or up to what point is a property of nature, or better to say, of dynamical systems in general (On this we call the attention of the reader to the recent book by P. Davies *The Fifth Miracle*, Simon and Schuster, New York, 1999). According to Jaynes, the question of how to apply theory of probabilities in science in theoretically valid and pragmatically useful forms was faced by Sir Harold Jeffreys [219, 223], in the sense that he stated a general philosophy of scientific inference and proceded to develop a mathematical theory and its implementations, as already noticed.

In MaxEnt-NESOM, the fact that a probability distribution, subject to certain constraints consisting in the information we possess, maximizes the informational-statistical entropy is a fundamental property that justifies the use of such distribution to perform inferences: it is in accord with the knowledge (information) we have, while carefully avoids to introduce any fact that we do not know. It has been said that it enforces the principle of economy in logic known as "Occam's razor": entities should not be multipied except of necessity. Supposedly, the MaxEnt-probability distribution should provide the best predictions that we can infer from the available information. As said, Predictive Statistical Mechanics is based on the Bayesian method in probability theory. On this, Philip W. Anderson has stated that these statistics are the correct way to do inductive reasoning from necessarily imperfect data; what Bayesianism does is to focus one's attention on the question one wants to ask of the data. [220].

The philosophy behind the theory is of course the object of lively, sometimes rispid, debate, and some considerations on that can be found in the book by Lawrence Sklar [1]. On practical grounds, one difficulty resides in how to decide on the basic set of macroscopic variables to be used, or, in other words, the choice of the informational space. For the case of mechanical systems this difficulty is in large part overcome resorting to the use of the selection rule provided by Zubarev-Peletminskii law of Eq. (2.4), which is, as noticed, the statistical mechanical counterpart of the principle of equipresence required in some phenomenological kinetic-thermodynamic theories. According to it one should introduce all dynamical quantities (micromechanical observables) that are quasi-conserved under the dynamics generated by the partial Hamiltonian H_0 of Eq. (2.3). The average values of them over the nonequilibrium ensemble are the macrovariables of Eq. (2.7) which decay with relaxation times larger than the characteristic time involved in the experiment. One may then say that we have a complete set of variables for the thermodynamic description of the system in the given *conditions fixed by the experimental situation under consideration.*

Once the basic set of dynamical quantities (micromechanical observables) has been decided upon, the statistical operator is built according to MaxEnt, but, we stress, with a particular choice of the Lagrange multipliers that the variational method introduces, in such a way to include memory effects in the evolution from an initial state of preparation, evolution which proceeds in an irreversible way. As noticed and discussed in Chapter 2, this implies in the introduction of a particular *Stosszahlansatz* (or Posit of Molecular Chaos). This is done in a peculiar form consisting in disregarding the subset of advanced solutions of Liouville equation, or introducing the so-called Boltzmann-Prigogine asymmetry. In such way the procedure belongs to the so-called *interventionist approaches* [1]. It is also noted by Sklar that several different approaches are related to one another, but exact or rigorous demonstrations of equivalences are rare. One case in hand is the relation of the here considered MaxEnt-NESOM and Prigogine and the Brussel's School method, which has been analyzed by J. P. Dougherty [224–226]. In his words, while the technical details of these approaches are both very formidable and quite different, it is possible to see in general terms how they are related. It is suggested that for many dynamical systems the approaches are likely to lead to the same results. In relation to MaxEnt it is also worth to cite Dougherty, in the sense that the fullfilments of the aims of statistical mechanics "can be regarded as a paradigm for science generally; and in that connection one may recall Popper's phrase '... *the art of discerning what we may with advantage omit*' [226]", meaning in MaxEnt-NESOM to retain the set of variables considered as relevant, but neglecting those that appear as irrelevants (of negligible influence on the characteristics of the system for which one is looking for), as we have described in previous chapters.

Sometimes it has been cited a kind of difficulty associated to the MaxEnt formalism, consisting in the meaning of the Lagrange multipliers. In the case of physical-chemical systems, that is in MaxEnt-NESOM, no such difficulty exists. As already noticed in Chapter 2 and thereafter, first, they are univocally and completely determined by the values of the macrovariables in IST, and second they acquire a well defined physical meaning once they are the differential coefficients of the IST-entropy, playing the role of intensive nonequilibrium thermodynamic variables, which, we stress, alternatively and equivalently, completely describe the thermodynamic state as the specific macrovariables do. Note also that, precisely, Eq. (3.3) implies in a Legendre-like transformation leading from the description in terms of the Q_j, on which $\bar{S}$ depends, to a description in terms of the F_j, on which ϕ depends which then can be considered as a nonequilibrium Massieu-Planck functional (we recall also its role as the logarithm of a nonequilibrium partition function). Moreover, because of these two points, we draw the attention to the interpretation we have done of some of the Lagrange multipliers that, typically, are presented as quasitemperatures, quasi-chemical potentials, quasipressure, quasi-drift velocities of densities and fluxes, etc.

Once the nonequilibrium statistical operator has been built, two important, and fundamental, following steps are in order: To derive a kinetic theory or nonlinear quantum generalized transport theory, and a response function theory. They follow, in principle, in a straightforward way, but, however, the mathematical manipulation is quite cumbersome and difficult, although a practical way for handling them in simplified conditions can be devised. The equations of evolution in the MaxEnt-NESOM kinetic theory, as shown, are simply the average over the nonequilibrium ensemble of the Heisenberg (or Hamilton) equations of evolution for the corresponding dynamical quantities (micromechanical observables). A response function theory and a scattering theory can also be derived by applying perturbation theory in the interaction representation, followed by averaging over the nonequilibrium ensemble.

On the basis of the background provided by this MaxEnt-NESOM one can study and analyze a large class of experimental situations, and provide relevant insight on the microscopic and macroscopic properties of biological, chemical, and condensed matter systems in general. Moreover, once the formalism allows to deal with nonlinear open dynamical system away from equilibrium, it provides — as noticed before — the substrate of a micro-dynamics from which to derive the macro-dynamics that can be used to analyze complex behavior, as synergetic selforganization, and also to build a nonlinear nonclassical thermo-hydrodynamics, and, as shown, the informational statistical thermodynamics we have considered here.

Concerning the latter, that is IST, we have dealt in detail in Chapters 3 and 4 with the properties of the entropy-like state function to it associated, that is the IST-entropy or informational entropy in IST. Let us add some additional consideration on the matter.

On entropy, Sklar [1] notes that the concept of entropy is the most purely thermodynamic concept of all, and Bricmont [227] has commented that there is some kind of mystique about entropy. As known, the concept originally arised out of the method introduced by Clausius of proving entropy's existence as a state function from the basic consequences of the second law. Again according to Sklar, given the abstractness of entropy and its high place up in the theory as well as its unrelatedness to immediate sensory qualities or primitive measurements (as temperature is related to these), it is not surprising than in seeking the statistical mechanical correlate of nonequilibrium thermodynamic entropy we have the least guidance from the surrounding embedding theory. Exists an openess in what to choose as the surrogate for entropy. Thus, it should be expected to arise a wide variety of "entropies", each functioning well for the specific purposes for which it was introduced. J. Meixner [228] asks: *Is the concept of nonequilibrium entropy superfluous?*, for in continuation to comment that one is so much accustomed to the concept of entropy that one would like to retain it as a quantity of physical significance. He also points to the difficulty of a definition if one does not have a clearly defined physical state of the system. This is the main difficulty, as also pointed out by Bricmont in that we may define as many entropies as we can find sets of macroscopic variables. Also, with the coarse-graining procedure there is not a sharp distinction between microscopy and macroscopy, passing through mesoscopy, so that we can arrive at many values of the say "entropy", including arriving at the zero value when a complete microscopic discription is given, and we have a pure mechanical description and no thermodynamics exists ([228, 229], and also Jaynes in [67]). Jaynes rightly says that he does not know what is the entropy of a cat; the problem being that we are unable to precisely define a set of macrovariables that properly specify the thermodynamic state of the cat.

At this point it is worth to present, with some modifications, several quite appropriate remarks made by Bricmont in the cited reference, which we roughly summarize here:

(i) These entropies are not subjective but objective as are the corresponding macroscopic variables. Called "anthropomorphic" by Jaynes following Wigner, they may be referred to as "contextual", i.e. they depend on the physical situation and on its level of description.

(ii) The "usual" or "traditional" entropy of Clausius corresponds to the particular choice of macroscopic variables for a free monoatomic gas in equilibrium (energy, specific volume, number of particles). The derivative with respect to the energy of *that* entropy defines the reciprocal of the Kelvin's absolute temperature.

(iii) The second law in the form "Entropy increases" becomes undetermined: which entropy? Between two states of equilibrium is the Clausius' one. Otherwise is not clear.

(iv) Whichever the chosen functional form for an entropy, in most cases is hard to compute or estimate. One needs to begin with the equations of evolution for the chosen basic variables and solve them for appropriate initial and boundary conditions. Moreover, irreversibility as characterized by some $\mathcal{H}$-theorem — as the one of Boltzmann — does not directly relate the $\mathcal{H}$-function to whatever may be the entropy. It only ensures that the choice of the initial condition and some ad hoc nonmechanical hypotheses (a Stosszahlanzatz) gives a time-arrow and relaxation towards final equilibrium.

(v) There is no difficulty with Liouville theorem of invariance of extension in phase space; this is a purely mechanical result. At the statistical level the extent of the volume of space points, compatible with the macroscopic description, changes because these constraints change in time, and the characteristic set of microscopic points changes. The evolution of such set is a different thing that the set of trajectories of given points in a volume of phase space, whose volume is indeed conserved according to Liouville theorem. Moreover,

(vi) we have noticed in the derivations in Chapter 4, that Gibbs' entropy is in fact constant in time, because MaxEnt-NESOM conserves the initial information, being then the said fine-grained entropy. But, as Fig. 2.1 shows, as time elapses it gets outside the informational subspace and therefore is no longer describing the state of the system in terms of the chosen set of basic variables, thus introducing loss of information. It needs be projected at each time on the informational subspace, again as shown in Fig. 2.1. Of course equilibrium is a particular case when Gibbs' entropy coincides with Clausius thermodynamic entropy, and in MaxEnt-NESOM it remains always at the point of initial preparation (which is that of the equilibrium).

(vii) Bricmont makes a similar statement to that of Meixner: Why should one worry so much about entropy for nonequilibrium states? To account for the irreversible behavior of the macroscopic variables is not necessary to introduce some kind of entropy function that evolves monotonically in time. It is not required to account for irreversibility, however it may be interesting or useful to do so. In the case we have presented of a MaxEnt-NESOM informational entropy applies this question of interest and usefulness: it allows for a better clarification of the meaning and interpretation of the Lagrange multipliers; to introduce the production of informational entropy and the derivation of useful criteria for evolution and stability; to better characterize the dissipative processes that develops in the media (organizing them in increasing orders of covariances of the informational-entropy production operator); to better analize fluctuations out of equilibrium and work out studies on complementarity of micro/macro descriptions; etc.

Therefore, in principle and to all appearances, a true thermodynamic entropy is only clearly defined, via Clausius approach, in strictly equilibrium conditions. Out of equilibrium, quasi-entropies (in our nomenclature) may be introduced and be of utility, but it needs be clearly stated which is the definition, and to receive a particular name that characterizes it, say entropy in Classical (Onsagerian) Irreversible Thrmodynamics, entropy in Extended Irreversible Thermodynamics, and the here presented, IST-entropy or MaxEnt-NESOM informational entropy, and so on and so forth.

In conclusion, the MaxEnt-NESOM, that is, the formulation described in Chapter 2, constitutes a soundly based, very poweful, concise, and practical mechanical-statistical formulation, which provides microscopic foundations for a nonlinear irreversible thermodynamics. The thermodynamical variables are introduced as constraints in the variational procedure (MaxEnt) for the derivation of the nonequilibrium statistical operator, which then depends on these variables through the Lagrange multipliers and the corresponding dynamical quantities (micromechanical observables).

The informational-entropy, to be related to the one in phenomenological theories, depends on these variables and arises from a projected part of the logarithm of the said nonequilibrium statistical operator $\ln \varrho_\varepsilon(t)$, which is $\ln \bar{\varrho}(t,0)$ (as shown in Fig. 2.1). This process is a coarse-grained-type procedure which restricts us to have as only accessible microstates in phase space those in the subspace spanned by the basic dynamic variables generated by the selection rule of Eq. (2.4). Irreversible effects are contained in the complementary part of the nonequilibrium statistical operator, namely ϱ'_ε [cf. Eq. (2.17)], as it is demonstrated by the proven generalized $\mathcal{H}$-theorem.

Furthermore, the local informational-entropy production function is predominantly positive definite, but in any case these results can be connected with the formulation of the second law, until a clear cut definition of the entropy function in nonlinear thermodynamics for systems arbitrarily away from equilibrium is obtained. This statement has to be further clarified since here the second law must necessarily be understood as some extension of Clausius formulation, with the latter, we recall, involving changes between two equilibrium states and S is the calorimetric entropy which is uniquely defined.

We emphasize once more that in nonequilibium states it is very likely that many different definitions of surrogates of the entropy are feasible depending essentially on how to obtain — in some sense — a complete set of macrovariables that may unequivocally characterize the macrostate of the system under the given experimental conditions, and thus agreeing with Meixner conjecture [228]. We stress the point, on which we have already commented upon, that in MaxEnt-NESOM an approximate complete set of thermodynamic variables can be obtained once an appropriate criterion for the truncation procedure in the choice of these basic variables can be develop in each particular problem under consideration.

Summarizing, it may be stated that MaxEnt-NESOM seems to offer a sound formalism to found irreversible thermodynamics on a statistical-mechanical basis, an approach providing what has been referred-to as Informational Statistical Thermodynamics. Thus, for the case of systems under quite arbitrary dissipative conditions (no restriction to local equilibrium, linearity, etc.) a theoretical treatment of a very large scope follows for the thermodynamics, transport properties, and response functions of nonequilibrium systems. Paraphrasing Zwanzig [90], we remark that, seemingly, the MaxEnt-NESOM possesses a remarkable compactness and has by far a most appealing structure, being a very effective method for dealing with nonlinear and nonlocal in space and time transport processes in far-from-equilibrium many-body systems.

APPENDIX I: Chronology of Relevant Events in the History of Thermodynamics and Statistical Mechanics

400BC: Leucippus of Miletos and Democritus of Abdera seem to have been the first to suggest that matter is composed of diminute particles – they are the first atomists.
100BC: Lucretius, the Roman poet, retake the atomistic idea (incorporated, in the V century BC to the IV century BC, by the philosopher Epicurus in a materialistic philosophy of life) in his book *De Rerum Natura.*
Medieval Times: Atomism is practically ignored in the Occidental world, and receiving scarce attention in the Muslim world.
~ 1592: **Galileo Galilei** develops a thermoscope.
1620 : **Francis Bacon, Baron Verulam**, publishes *Instauratio Magna*, presenting a proposal for a comprehensive reorganization of the sciences, an important contribution to the so-called Scientific Revolution of the XVII century.
~ 1632: **Jean Ray** develops a thermometer based on the expansion of liquids (open capillar) [There seems to be a precursor in **Sanctorius Sanctorius** (1611), a colleague of Galileo].
1634: **Galileo** completes his most valuable work on mechanics, *Discorsi e dimostrazioni matematiche intorno a due nove scienze attenenti alla mecanica* (usually reffered to as *Dialogue Concerning Two New Sciences*), printed in Leiden in 1638.

~ **1640: Duke Ferdinand II of Tuscany** develops an alcohol thermometer (of close capillar).

~ **1640: Bacon** advances the idea that it is not temperature that is transmitted from hot to cold bodies.

~ **1661: Robert Boyle** ponders the idea of chemical element.

~ **1660: Robert Hooke** considers concepts that are close to what later on will be kinetic theory.

1687: Isaac Newton publishes his monumental *Philosophiae Naturalis Principia Mathematica.*

1716: J. Hermann suggests that heat is due to the movement of particles.

1727: Leonard Euler iniciates kinetic theory: he propounds that air consists of particles, develops a theory of humidity, and notes that pressure and temperature are macroscopic manifestations of molecular (particle) action.

1738: Daniel Bernouilli replicates Euler's results in an extended form, and suggests the use of kinetic energy as a scale of temperature. Also, he forebodes the principle of energy conservation including heat, with the latter being a manifestation of atomic movement.

~ **1759: Thomas Bayes** writes the "chain rule" for probability theory (The paper was published posthumously in 1763) He is the first to use probability inductively and to establish a mathematical basis for probability inference, giving rise to what is now called Bayesian Statistics.

1770: Joseph Black reinforces the distinction between temperature and heat and gives rise to the science of calorimetry, leading to the caloric theory.

~ **1770: Ruggero Boscovich** suggests the atoms as being not only simply diminute particles but centers of forces.

1782: Leonard Euler extends Bernouilli's ideas, attempting the first serious efforts to replace phenomenological temperature with definitions based on atomic movement.

1798: Benjamin Thompson, Count Rumford, suggests the equivalence of heat and work, reinforcing the idea that the former is a manifestation of the movement of particles. He criticizes severely the caloric theory, whose end is in sight.

1807: Jean-Baptiste-Joseph Fourier establishes the theory of transmission of heat.

1808: John Dalton proposes on strong grounds the atomistic theory in his book *A New System of Chemical Philosophy.*

1811: Amedeo Avogadro introduces the concept of molecules (as bonded aggregates of atoms), and enunciates Avogadro's principle.
1814: Pierre Simon, Marquis de Laplace introduces the concepts of probability theory in Physics.
1816: Laplace gives a correct adiabatic treatment of the propagation of sound.
1821: J. Herapath provides a rudimentary kinetic theory, in a tentative explanation of changes of state, diffusion, and sound propagation.
1824: Nicolas Leonard Sadi-Carnot perceives what will be the second law: any thermal engine cannot be more efficient than a reversible one. He introduces the concept of thermodynamic cycles, what permits a clear distinction between interactions in a system and its changes of state.
1825: M. Seguin presents a careful discussion on the relation between heat and molecular movement.
1840: James Prescott Joule in the period 1840–1849, performs a series of experiments that determine the mechanical equivalent of heat, and establishes the fundamentals of the first law.
1842: Julius Robert von Mayer reinforces the relevance of the conservation of energy in all its forms.
1843: J. J. Waterston provides a complete kinetic theory , deriving that $p = \langle nv^2 \rangle /3$, and noticing that temperature is proportional to some kind of average of the square of the velocity. He hinted an equipartition law. This is the first viable kinetic theory.
1844: Mayer asserts that the heat introduced in a cycle exceds the work done.
1847: Hermann Ludwig Ferdinand von Helmholtz enunciates the equivalent of the first law of thermodynamics.
1848: William Thompson, Lord Kelvin, introduces the thermometric scale that bears his name.
1850: Rudolf Julius Clausius makes explicit the first and second laws of Thermodynamics, which then arises as a well defined science.
1851: Joule rederives Waterston's work, but without mentioning average values, and reobtains the expression for the pressure.
1853: Kelvin expands and improves upon the work of Joule of 1851, as well as on that of Waterston and firmly establishes $p = \langle nv^2 \rangle /3$.
1856: A. K. Krönig reviews and summarizes the state of the art, at that time, of kinetic theory. Although he does not add any new idea, his prestige at the time lends a good support to the theory.

1857: Clausius publishes work initiated around 1850 making more specific the ideas that give ground to thermodynamics, and distinguishes the three different states of matter in terms of its molecular properties. At this stage caloric theory is fadding away, and kinetic theory acquires relevance, however it is not universally accepted.
1858: Clausius introduces for the first time the explicit notion of probability in kinetic theory, defining what needs be understood by average value. He defines the mean free path, making clear the difference between it and the average intermolecular spacing $n^{-1/3}$. He perceives the necessity of some type of Stosszahlansatz, that is, the assumption of molecular chaos referring to the statistical independence of the molecules prior to a collision, but does not formulate it.
1859: Gustav R. Kirchhoff and Robert W. Bunsen establish the principles of spectroscopy.
1860: James Clerk Maxwell develops the modern vision of kinetic theory. He derives the velocity distribution for point particles, and finds an expression for the mean free path in terms of the temperature, the mass and the radius of the "sphere of influence" of the molecule. He predicts that the viscosity in a gas is independent of the density, the first prediction of a property using a molecular model.
1864: Maxwell publishes his famous treatise A Dynamical Theory of Electromagnetic Field.
1865: Clausius defines, and names, the concept of entropy. He publishes his book on the Theory of Heat pointing out to the probabilistic nature of the second law.
1865: Joseph Loschmidt uses kinetic theory to estimate that the radius of the "sphere of influence" of molecules is of the order of 10^{-8} cm.
1866: Ludwig Boltzmann publishes his first article on statistical mechanics, and states his objective to derive the first and and second law of Thermodynamics on purely mechanical concepts.
1867: Maxwell publishes a new version of kinetic theory: he resolve certain difficulties inherent to the previous approach, and founds a well structured physico-mathematical kinetic theory. In the same year introduces the concept of the, presently named, "Maxwell's Demon", and emphasizes the probabilistic character of the second law.
1867: Maxwell, in the theory on viscosity, argues that in a viscous media the state of strain would tend to disappear at a rate that depends on the state of stress and the nature of the body. (This idea will be, in

the second half of 20th century, carried ahead by Extended Irreversible Thermodynamics).

1868: Boltzmann extends kinetic theory so as to include Maxwell-Boltzmann distribution, and also allowing for the cases of molecules and external fields.

1869: Dimitri Mendeleiev builds the periodic table of the elements.

1870: Kelvin iniciates a discussion on the size of atoms.

1870: Clausius develops the virial theorem.

1872: Boltzmann introduces his famous equation, incorporating the Stosszahlansatz, and the $\mathcal{H}$-theorem is born.

1877: Boltzmann emphasizes the probabilistic nature of the second law, in accord with Maxwell, and introduces the method of the most probable values. He expresses the idea of the famous relation $S = k_B \ln W$, but does not write it explicitly, what is done by Planck later on.

1878: Josiah Willard Gibbs publishes his notable work *On the Equilibrium of Heterogeneous Systems.*

1879: William Crookes finds indications of the existence of negatively charged particles in experiments on cathodic rays.

1887: Heinrich Hertz produces and detects electromagnetic waves in laboratory.

1896: Boltzmann publishes his famous book *Vorlesungen über Gastheorie.*

1897: Joseph John Thompson establishes the existence of the electron.

1900: Max Planck derives the distribution function that bears his name, to solve the enigmatic question of the spectral distribution of the energy irradiated by the black body, in a scheme that implies the introduction of the quantum of radiation energy, and makes to appear the famous Planck constant.

1902: Gibbs publishes the fundamental book *Elementary Principles in Statistical Mechanics*, defining and given the foundations for this emergent science.

1905: Albert Einstein proposes a quantitative theory of the Brownian motion.

1906: Planck gives form to Boltzmann's idea that $S = k_B \ln W$, where k_B is Boltzmann constant and W is the number of microscopic mechanical states (or volume in phase space in the classical case) compatible with the macroscopic state of the system.

1908: Jean-Baptiste Perrin verifies Einstein theory, and reinforces the atomistic view, with his experiments on Brownian motion.

1909: Constantin Caratheodory presents an alternative logical structure to Thermodynamics, trying to avoid the concept of heat.

1910: Lord Rutherford proposes the nuclear model of the atom.

1911: Paul and Tatiana Ehrenfest publish their influencial article on the foundations of Statistical Mechanics.

1913: Niels Bohr presents the stationary planetary model of the atom.

1918: Walther Nernst establishes the third law of Thermodynamics.

$\gtrsim$ 1920: Beginning and development of **Quantum Mechanics**: Louis de Broglie (1923); Niels Bohr, Werner Heisenberg, Max Born, Pascual Jordan and others in the The Copenhagen School (from nearly 1920 on); Erwin Schrödinger (1926); Paul Dirac (1929), etc.

1922: Otto Stern and Walter Gerlach perform an experiment on the deflection of atomic beams in magnetic fields.

1925: George Uhlenbeck and Samuel Goudsmit incorporate the idea of intrinsic angular momentum, or spin, and derive its consequences.

1926: Satyendra Nath Bose and Albert Einstein develop the distribution that bears their names.

1927: Enrico Fermi and Paul M. Dirac develop the distribution that bears their names.

1929: Wolfgang Pauli publishes the first quantum mechanical version of the master equation.

1930: Johannes von Newmann and Paul M. Dirac introduce the density matrix in quantum mechanics, which is of later relevance in Statistical Mechanics.

1931: Lars Onsager establishes the foundations of **Classical Irreversible Thermodynamics.**

1931: Harold Jeffreys collects in a book a series of previous papers on the question of logic and scientific inference in science and knowledge in general. The points of view there expressed are to give a framework for later developments in informational statistical mechanics and thermodynamics.

1935: J. Yvon gives a first systematic treatment of the reduced distribution functions, origin of the so called BBGKY hierarchy of equations.

1939: Jeffreys publishes his book *Theory of Probability,* dealing with the question of inference in sciences and the role of Bayesian Statistics.

1941: J. Meixner provides a general and comprehensive treatment of Classical Irreversible Thermodynamics.

1945: H. B. G. Casimir publishes a comprehensive review article on Onsager's principle of microscopic reversibility.

1945: Ilya Prigogine generalizes Classical Irreversible Thermodynamics, and introduces the theorem of minimum entropy production.

1946: Nicolai N. Bogoliubov presents a systematic introduction of the functional methods in statistical mechanics, and gives the most rigorous form of the BBGKY hierarchy.

1946: John G. Kirkwood introduces a mechano-statistical approach in transport theory.

1948: Claude Shannon introduces Information Theory, which seems to generalize the concept of entropy.

1951: Herbert B. Callen and T. A. Welton derive the general form of the fluctuation-dissipation theorem.

1951: Sveig R. de Groot publishes *Thermodynamic of Irreversible Processes*, a compendium and critique of applications of the Onsager reciprocity theorem.

1952: Melvin S. Green introduces an statistical approach to irreversible processes, one of the first attempts to create a nonequilibrium ensemble theory.

1954: Leon van Hove introduces, in linear response theory, the relation between the time correlation and the inelastic scattering cross section.

1956: Peter T. Landsberg introduces a conceptual extension of Caratheodory's method in a geometrical-topological approach.

1957: Ryogo Kubo develops a complete theory of dynamical response and transport phenomena.

1957: Edwin T. Jaynes elaborates an in depth analysis of Gibbs' approach in a Statistical Mechanics related to the point of view of Information Theory. He introduces the Principle of Maximization of (informational-statistical) Entropy, or MaxEnt, retrieving Laplace's ideas in probability theory, and Jeffreys' scientific inference point of view.

1958: Bogoliubov and Sergei V. Tyablikov complete the formalism of the double-time thermodynamic Green functions [see in Bibliography, Zubarev (1960)].

1961: Robert Zwanzig introduces the use of projection operators in Statistical Mechanics.

1962: Bogoliubov states the principle of correlation weakening and the hierarchy of relaxation times, seemingly fundamental to the development of reliable methods of nonequilibrium statistical thermodynamics.
1962: de Groot and P. Mazur publish their, nowadays classical, book on linear irreversible thermodynamics.
1963: Prigogine introduces the idea of dynamics of correlations, rather than that of distribution functions of the BBGKY hierarchy, for dealing with nonequilibrium statistical mechanics of systems far from equilibrium.
1963: Beginning of **Chaos Theory**, starting from the motivation of the articles by E. N. Lorenz (The concept goes back to the studies of Henry Poincaré, (~1890–99) on stability problems in celestial mechanics).
1965: Hazime Mori develops a theory of generalized statistical Newton-Langevin equations.
~ **1965**: The development of **Extended Irreversible Thermodynamics** begins at the hands of several authors [among others: R. Nettleton (1960); I. Müller (1967); G. Lebon (1973); I. Gyarmati (1977); J. Casas-Vázquez and D. Jou (1979,1988)].
1966: A. Hobson advances the, apparently, first attempt to build a statistical thermodynamics on the foundations of the 1957-Jaynes' first approach to statistical mechanics on the basis of information theory: this appears as providing the initiation of **Informational Statistical Thermodynamics**.
1966: Leo P. Kadanoff advances the hypothesis of scaling in the theory of critical phenomena.
1969: C. Truesdell publishes his book on **Rational Thermodynamics**.
1969: Prigogine conjectures on the possible connection between **dissipative structures** (i.e. complex behavior in open systems far from equilibrium, that he evidenced) and the emergence, functioning and evolution of life.
1971: Dimitri N. Zubarev presents his formulation of an approach to the nonequilibrium statistical operator method.
1971: P. Glansdorff and I. Prigogine publish their book on **Generalized Classical Irreversible Thermodynamics** extended to the nonlinear domain. It includes the thermodynamic principles of evolution and (in)stability in systems under arbitrary nonequilibrium conditions.
1971: Kenneth Wilson gives a compact form to the theory of critical phenomena in terms of the Renormalization Group.

1972: Philip W. Anderson publishes what is considered one of the first "manifestos" referring to a seemingly new paradigm in sciences consisting in the so-called **Theory of Complexity**.
1975: Benoit Mandelbrot introduces fractal geometry, which has relevant role in the theory of dynamic systems.
1979: Prigogine, in collaboration with **Isabelle Stengers**, publish *La Nouvelle Alliance: Metàmorphose de la Science*, with philosophical considerations on the emergence of order (self-organization) in dynamical systems.
1980: Prigogine publishes his book *From Being to Becoming* on time (irreversibility) and complexity (self-organization) in physical sciences.
~ **1983: Jaynes** reinforces his proposal of Predictive Statistical Mechanics, containing MaxEnt, and based on Bayesian methods of probability.
1990: Several authors are proceeding to extend Hobson's original pioneering work on **Informational Statistical Thermodynamics**, a discipline that appears to relate Jaynes' Predictive Statistical Mechanics and phenomenological Irreversible Thermodynamics, , the one which has received some attention in the present book.

Bibliography to the Appendix I

The names cited in the chronology above are listed in continuation in alphabetical order; those cited in the chronology and not listed below is due to the fact that we have not been able to trace the reference of the original work.

ANDERSON, P. W. (1972) 'More is different', *Science* **177**, 393.

BAYES, T. (1763) 'Essay Towards Solving a Problem in the Doctrine of Chances', published post humously in *Phil. Trans. Roy. Soc. (London)* pp. 330–418.

BERNOUILLI, D. (1738) *Hydrodynamicas.* Argentorati.

BOGOLIUBOV, N. N. (1962), in *Studies in Statistical Mechanics I,* J. Boer and G. E. Uhlenbeck, Eds. North Holland, Amsterdam.

BOLTZMANN, L. (1866) 'Über die mechanischen Bedeutung des zweiten Hauptsatzes der Wärmetheorie', *Wien. Ber.* **53**, 195.

BOLTZMANN, L. (1868) 'Studien über das Gleichgewicht der lebendigen Kraft zwischen bewegten materiellen Punkten', *Wien. Ber.* **58**, 517.

BOLTZMANN, L. (1871) 'Wärmetheorie aus den Sätzen über das Gleichgewicht der lebendigen Kraft', *Wien. Ber.* **B**, 712.

BOLTZMANN, L. (1872) 'Weitere studien über das Wärmegleichgewicht unter Gasmolekülen', *Wien. Ber.* **66**, 275.

BOLTZMANN, L. (1877a) 'Bemerkungen über einige Probleme der mechanischen Wärmetheorie', *Wien. Ber.* **75**, 62.

BOLTZMANN, L. (1877b) 'Über die Beziehung zwischen dem zweiten Hauptsatzes der mechanischen Wärmetheorie und der Wahrscheinlichkeitsrechnung respektive den Sätzen über das Wärmegleichgewicht', *Wien. Ber.* **76**, 373.

BOLTZMANN, L. (1887) 'Über die mechanischen Analogien des zweiten Hauptsatzes der Thermodynamik', *J. r. ang. Math.* **100**, 201.

BOLTZMANN, L. (1895) 'On Certain Questions of the Theory of Gases', *Nature* **51**, 413, 581.

BOLTZMANN, L. (1896) 'Entgegnung auf der Wärme theoretischen Betrachtungen des Hrn. E. Zermelo', *Wied. Ann.* **57**, 773.

BOLTZMANN, L. (1896) *Vorlesungen über Gastheorie.* Barth, Leipzig (Part I, 1896; Part II, 1898).

BOGOLIUBOV, N. N. (1946) *Problemy Dinamicheskoi Teorii Statisticheskoi Fisike.* Gostekhizdat, Moscow.

CALLEN, H. B., Welton, T. A. (1951) 'Irreversibility and General Noise', *Phys. Rev.* **83**, 34.

CARNOT, S. (1824) *Reflexions sur la puissance mortice du feu et sur les machines propres à developper cette puissances.* Bachelier, Paris.

CASIMIR, H. B. G. (1945) 'On Onsager's Principle of Microscopic Reversibility', *Rev. Mod. Phys.* **17**, 343.

CLAUSIUS, R. (1857) 'Über die Art der Bewegung, welche wir Wärme nennen', *Pogg. Ann.* **100**, 253.

CLAUSIUS, R. (1859) 'Über die mittlere Länge der Wege, welche bei der Molekularbewegung gasförmiger Körper von der einzelnen Molekülen zurückgelegt werden, nebst einigen anderen Bemerkungen über die mechanischen Wärmetheorie', *Pogg. Ann.* **105**, 239.

CLAUSIUS, R. (1865) 'Über verschiedene für die Anwendung bequeme Formen der Hauptgleichungen der mechanische Wärmetheorie', *Ann. d. Phys.* [2] **125**, 390.

CLAUSIUS, R. (1870) 'Über einen auf die Wärme anwendbaren mechanischen Satz', *Ann. d. Phys.* [2] **141**, 124.

DE GROOT, S. R. (1951) *Thermodynamics of Irreversible Processes,* North Holland, Amsterdam.

DE GROOT, S. R., Mazur, P. (1962) *Non-Equilibrium Thermodynamics,* North Holland, Amsterdam.

EHRENFEST, P. and EHRENFEST, T. (1911) 'Begriffiche Grundlagen der statistischen Auffassung in der Mechanik', Vol. IV, Part 32, *Enzyklopädie der Mathematischen Wissenschaften.* Teubner, Leipzig [English version (1959): *The Conceptual Foundations of the Statistical Approach in Mechanics.* Cornell Univ. Press, Ithaca, N. Y.].

EINSTEIN, A. (1905) 'Zur Theorie der Brownschen Bewegung', *Ann. d. Phys.* **19**, 371.

EULER, L. (1727) 'Tentamen explicationis phænomenorum æris', *Comm. Acad. Sci. Petrop.* **2**, 347.

EULER, L. (1782) *Acta Acad. Sci. Petrop.* **1**, 162.

FOURIER, J. B. J. (1807) *Mémoire sur la Propagation de la Chaleur* (a summary was read in the session of the Institute de Paris in December 21, 1907).

GIBBS, J. W. (1876) 'On the Equilibrium of Heterogeneous Substances', *Trans. Conn. Acad.* **3,** 108, 343.

GIBBS, J. W. (1902) *Elementary Principles in Statistical Mechanics*, Yale Univ. Press, New Haven.

GLANSDORFF, P., PRIGOGINE, I. (1971) *Thermodynamic Theory of Structure, Stability, and Fluctuations*, Wiley-Interscience, New York.

GREEN, M. S. (1952) 'Markoff Random Processes and the Statistical Mechanics of Time-Dependent Phenomena', *J. Chem. Phys.* **20**, 1281 (1952); *ibid.* **22**, 398 (1954).

GYARMATI, I. (1977) 'On the Wave Approach of Thermodynamics and Some Problems of Non-Linear Theories', *J. Non-Equilib. Thermodyn.* **2**, 233.

VON HELMHOLTZ, H. (1847) *Über die Erhaltung der Kraft.* G. Reimer, Berlin.

HERAPATH, J. (1821) 'A Mathematical Inquiry into the Causes, Laws and Principal Phenomena of Heat, Gases, Gravitation, etc.', *Ann. Phil.* **1**, 273, 340, 401.

HERMANN, J. (1716) *Phoronomia sive de viribus et motibus corporum soidorum et fluidorum libri duo*, Amsterdam.

HOBSON, A. (1966) 'Irreversibility and information in mechanical systems', *J. Chem. Phys.* **45**, 1352.

VAN HOVE, L. (1954) 'Correlations in Space and Time and Born Approximation Scattering in Systems of Interacting Particles', *Phys. Rev.* **95**, 249.

JAYNES, E. T. (1957) 'Information Theory and Statistical Mechanics', *Phys. Rev.* **106**, 620.

JAYNES, E. T. (1983) *E. T. Jaynes' Papers on Probability, Statistics, and Statistical Physics*, R. D. Rosenkrantz, Ed. Reidel. Dordrecht-Holland.

JAYNES, E. T. (1985) 'Macroscopic Prediction', in *Complex Systems: Operational Approaches*, H. Haken, Ed. Springer, Berlin.

JAYNES, E. T. (1986) 'Predictive Statistical Physics', in *Frontiers of Nonequilibrium Statistical Physics*, G. T. Moore e M. O. Scully, Eds. Plenum, New York.

JAYNES, E. T. (1986) 'Bayesian Methods: General Background', in *Maximum Entropy and Bayesian Methods*, J. H. Justice, Ed. Cambridge Univ. Press, Cambridge.

JAYNES, E. T. (1989) 'Clearing up Mysteries: The Original Goal', in *Maximum Entropy and Bayesian Methods*, J. Skilling, Ed. Kluwer, Dordrecht-Holland.

JAYNES, E. T. (1991) 'Notes on Present Status and Future Prospects', in *Maximum Entropy and Bayesian Methods*, W. T. Grandy e L. H. Schick, Eds. Kluwer, Dordrecht-Holland.

JEFFREYS, H. (1931) *Scientific Inference*, Cambridge Univ. Press, Cambridge.

JEFFREYS, H. (1939) *Theory of Probability*, Oxford Univ. Press, Oxford.

JOU, D., CASAS-VÁZQUEZ, J., LEBON, G. (1979) 'A dynamical interpretation of second-order constitutive equations of hydrodynamics', *J. Non-Equilib. Thermodyn.* **4**, 349.

JOU, D., CASAS-VÁZQUEZ, J., LEBON, G. (1988) 'Extended Irreversible Thermodynamics', *Rep. Prog. Phys.* **51**, 1105.

JOULE, J. P. (1845) In: Phil. Mag. **27**, 205.

JOULE, J. P. (1851) 'Some Remarks on Heat and the Constitution of Elastic Fluids', *Mem. Manchester Lit. Phil. Soc.* **9**, 107.

KADANOFF, L. P., et al. (1967) 'Static Phenomenon Near Critical Points', *Rev. Mod. Phys.* **39**, 395; also (1966) *Physics* **2**, 263.

KIRKWOOD, J. G. (1946) 'The Statistical Mechanical Theory of Transport Processes', *J. Chem. Phys.* 14, 180.

KRÖNIG, A. K. (1856) 'Grundzüge einer Theorie der Gase', *Pogg. Ann.* **99**, 315.

KUBO, R. (1957) 'Statistical Mechanical Theory of Irreversible Processes. I. General Theory and Simple Applications to Magnetic and Conduction Problems', *J. Phys. Soc. Japan* **12**, 570.

LAMBERMONT, J., LEBON, G. (1973) 'On a generalization of Gibbs equation for heat conduction', *Phys. Lett. A* **42**, 499.

LANDSBERG, P. T. (1956) Foundations of Thermodynamics, *Rev. Mod. Phys.* **28**, 363.

LANDSBERG, P. T. (1961) *Themodynamics with Quantum Statistical Illustrations.* Interscience, New York, 1961.

LAPLACE, P. S. (1816) *Ann. Phys. Chim.* **3**, 288.

LAPLACE, P. S. (1825) *Essai philosophique sur les probabilités*, 5th ed., Bachelier, Paris.

MANDELBROT, B. (1983) *The fractal geometry of nature*, Freeman, New York.

MAXWELL, J. C. (1859) 'Letter to G. G. Stokes on 30 May', in *Memoir and Scientific Correspondence of the late George Gabriel Stokes*, Vol. 2, J. Larmor, Ed., Cambridge Univ. Press, Cambridge.

MAXWELL, J. C. (1860) 'Illustrations of the Dynamical Theory of Gases', *Phil. Mag.* **19**, 19, *ibid.* **20**, 21; *ibid.* **20**, 33.

MAXWELL, J. C. (1867) 'On the Dynamical Theory of Gases', *Phil. Trans. Roy. Soc. London* **157**, 49.

MAXWELL, J. C. (1867) 'Letter to P. G. Tait on 11 December', in C. G. Knott, *Life and Scientific Work of Peter Gutherie Tait*, C. G. Knott, Ed. Cambdrige Univ. Press, Cambridge, 1911.

MAXWELL, J. C. (1870) 'Letter to J. W. Strutt on 6 December', in *Life of John William Strutt, Third Baron Rayleigh*, Univ. Wisconsin Press, Madison 1968.

MAYER, J. R. (1842) 'Bemerkungen über die Kräfte der unbelebten Natur', *Ann. Chemie und Pharmacie* **42**, 233.

MORI, H. (1965) 'Transport, Collective Motion, and Brownian Motion', *Prog. Theor. Phys.* (Japan) **33**, 423.

MÜLLER, I. (1967) 'Zum Paradox der Wärmerleitungstheorie', *Zeit. Phys.* **198**, 329.

MÜLLER, I. (1985) *Thermodynamics*, Pitman, London.

NERNST, W. (1918) *Die teoretischen und experimentellen Grundlagen des neuen Wärmetheorems*, Knappe, Halle.

NETTLETON, R. (1960) 'Relaxation Thermal Theory of Conduction in Liquids', *Phys. Fluids* **3**, 216.

NETTLETON, R. (1990) 'Generalized Grad-type foundations for nonlinear extended thermodynamics', *Phys. Rev. A* **42**, 4622.

VON NEWMANN, J. (1932) *Mathematische Grundlagen der Quantenmechanik*, Berlin.

ONSAGER, L. (1931) 'Reciprocal Relations in Irreversible Processes I', *Phys. Rev.* **37**, 405; II, *ibid.* **38**, 2265.

PERRIN, J. (1916) *Atoms*, Van Nostrand, Princeton.

PLANCK, M. (1906) *Vorlesungen über die Theorie der Wärmestrahlung*, J. A. Barth, Leipzig.

PRIGOGINE, I. (1945) in *Bull. Classe Sci. Acad. Roy. Belg.* 31, 600; in detail in the treatise (1947) *Etude Thermodinamique des Phenomenes Irre-*

versibles, Desoer, Liége [also (1955) *Introduction to the Thermodynamics of Irreversible Processes.* Thomas, New York).
PRIGOGINE, I. (1963) *Non-Equilibrium Statistical Mechanics.* Wiley-Interscience, New York.
PRIGOGINE, I. (1969) 'Structure, Dissipation, and Life', in *From Theoretical Physics to Biology,* M. Marois, Ed. North Holland, Amsterdam.
PRIGOGINE, I., STENGERS, I. (1979) *La Nouvelle Alliance: Metamorphose de la Science,* Gallimard, Paris [extended English version with a Preface by A. Toffler (1984); *Order Out of Chaos: Man's New Dialogue with Nature,* Bantam Books, New York].
PRIGOGINE, I. (1980) *From Being to Becoming.* Freeman, San Francisco.
SEGUIN, M. (1825) 'Letter to Dr. Brewster on the Effects of Heat and Motion', *Edinburgh J. Sci.* **3**, 276.
SHANNON, C. E. (1948) 'A mathematical Theory of Communication', *Bell. System Tech. J.* **27**, 379, 623.
THOMPSON, B. (Count Rumford) (1798) In: *Phil. Trans. Roy. Soc. London* **80.**
THOMPSON, W. (Lord Kelvin) (1853) 'On the Mechanical Action of Heat, and the Specific Heats of Air', *Cambridge and Dublin Math. J.* **96**, 270.
THOMPSON, W. (1870) 'The Size of Atoms', *Nature* **1**, 551.
TRUESDELL, C. (1969) Rational Thermodynamics, McGraw-Hill, New York.
WATERSTON, J. J. (1893) 'On the Physics of Media that are Composed of Free and Perfectly Elastic Molecules in a State of Motion', *Phil. Trans. Roy. Soc. London* **183A**, 79. [Published posthumously; first submitted in 1843; abstract published in *Proc. Roy. Soc. (London)* **5**, 604 (1846)].
WILSON, K. (1971) 'Renormalization Group and Critical Phenomena', *Phys. Rev. B* **4**, 3174; *ibid.* **4**, 3184.
YVON, J. (1935) *La Théorie Statistique des Fluides et l'Equation d'Etat,* Hermann, Paris.
ZUBAREV, D. N. (1960) 'Double-Time Green Functions in Statistical Physics', *Usp. Fiz. Nauk* **71**, 71 [*Soviet Phys. — Uspekhi* **3**, 320 (1960)].
ZUBAREV, D. N. (1971) *Neravnovesnaia Statisticheskaia Termodinamika,* Izd. Nauka, Moscow [In English: (1974) *Nonequilibrium Statistical Thermodynamics,* Plenum-Consultants Bureau, New York].
ZWANZIG, R. (1961) 'Statistical Mechanics of Irreversibility', in *Lectures in Theoretical Physics, Vol. 3,* W. E. Brittin, B. W. Downs, and J. Davis, Eds. Wiley-Interscience, New York.

APPENDIX II: Production of Informational Entropy

According to Eqs. (3.10) and (3.11) the global informational entropy production, after we ignore the dependence on the space coordinate $\boldsymbol{r}$, is given by

$$\bar{\sigma}(t) = \sum_j F_j(t) \frac{d}{dt} Q_j(t) . \tag{II.1}$$

But

$$\frac{d}{dt} Q_j(t) = J_j^{(0)}(t) + J_j^{(1)}(t) + \mathcal{J}_j(t) \tag{II.2}$$

[cf. Eq. (2.24)]. Let us consider the influence of the first two terms on the right of Eq. (II.2), namely

$$\begin{aligned} \sum_j F_j(t) \left\{ J_j^{(0)}(t) + J_j^{(1)}(t) \right\} &= \sum_j F_j(t) \operatorname{Tr} \left\{ \frac{1}{i\hbar} \left[\hat{P}_j, \hat{H}_0 + \hat{H}' \right] \bar{\varrho}(t,0) \right\} \\ &= \operatorname{Tr} \left\{ \frac{1}{i\hbar} \left[\sum_j F_j(t) \hat{P}_j, \hat{H}_0 + \hat{H}' \right] \bar{\varrho}(t,0) \right\} . \end{aligned} \tag{II.3}$$

Recalling that

$$\phi(t) + \sum_j F_j(t) \hat{P}_j = \ln \bar{\varrho}(t,0) , \tag{II.4}$$

where $\phi(t)$ is a scalar function, then

$$\sum_j F_j(t) \left\{ J_j^{(0)}(t) + J_j^{(1)}(t) \right\} =$$

$$\mathrm{Tr}\left\{ \frac{1}{i\hbar} \left[\ln \bar{\varrho}(t,0), \hat{H}_0 + \hat{H}' \right] \bar{\varrho}(t,0) \right\} =$$

$$= \mathrm{Tr}\left\{ \frac{1}{i\hbar} (\hat{H}_0 + \hat{H}') \left[\ln \bar{\varrho}(t,0), \bar{\varrho}(t,0) \right] \right\} = 0 \,, \quad \text{(II.5)}$$

where we have used the invariance of the trace operator by cyclic permutation of the operators. Hence

$$\bar{\sigma}(t) = \sum_j F_j(t) \mathcal{J}_j(t) = \sum_j F_j(t)\, \mathrm{Tr}\left\{ \frac{1}{i\hbar} \left[\hat{P}_j, \hat{H}' \right] \varrho'_\varepsilon(t) \right\} \,, \qquad \text{(II.6)}$$

that is, the production of global informational entropy is generated by the collision processes present in $\hat{H}'$, and dissipation, as noted in the main text, is accounted for the contribution ϱ'_ε to the nonequilibrium statistical operator [cf. Eqs. (2.3) and (2.17)].

Bibliography

[1] L. Sklar, *Physics and Chance: Philosophical Issues in the Foundations of Statistical Mechanics*, Cambridge Univ. Press, Cambridge, 1993.

[2] J. Bronowski, *The Common Sense of Science*, Harvard Univ. Press, Cambridge, MA, 1978.

[3] R. P. Feynman, *The Character of Physical Laws*, The MIT Press, Cambridge, Ma, 1967, 1965 The British Broadcasting Corporation.

[4] W. Ebeling and W. Muschik, editors, *Statistical Physics and Thermodynamics of Nonlinear Nonequilibrium Systems*, World Scientific, Singapore, 1993.

[5] S. C. Brown, The Caloric Theory of Heat, Am. J. Phys. **18**, 367 (1950).

[6] P. S. Laplace, *Essay Philosophique sur les Probabilities*, Bachelier, Paris, 1825.

[7] P. W. Anderson, More is different, Science **117**, 393 (1972).

[8] P. W. Anderson, Is Complexity Physics? Is it Science? What is it?, Phys. Today **44**(7), 9 (July 1991).

[9] M. D. Lemonick, Life, the Universe, and Everything, Time Magazine, 40 (February 22 1993).

[10] R. Luzzi and A. R. Vasconcellos, Statistical Mechanics of Dissipation and Order, Ciência e Cultura (A publication of the Brazilian Society for the Advancement of Science) **43**, 423 (1991).

[11] E. T. Jaynes, Foundations of Probability Theory and Statistical Mechanics, in *Delaware Seminar in the Foundations of Physics*, edited by M. Bunge, Springer, New York, 1967.

[12] E. T. Jaynes, Bayesian Methods: General Background, in *Maximum Entropy and Bayesian Methods*, edited by J. H. Justice, Cambridge Univ. Press, Cambridge, 1986.

[13] E. T. Jaynes, On the Rationale of Maximum Entropy Methods, Proc. IEEE **70**, 939 (1982).

[14] L. S. García-Colín, A. R. Vasconcellos, and R. Luzzi, On Informational Statistical Thermodynamics, J. Non-Equilib. Thermodyn. **19**, 24 (1994).

[15] M. Planck, *Treatise on Thermodynamics*, Dover, New York, 1945, Translated from the seventh German edition; the quoted sentences are from the first edition in 1897.

[16] N. Georgescu-Roegen, *The Entropy and the Economic Process*, Harvard Univ. Press, Cambridge, Massachusetts, 1971.

[17] I. Prigogine, *Étude Thermodinamique des Phénomènes Irreversibles*, Desoer, Liège, 1947.

[18] S. de Groot and P. Mazur, *Nonequilibrium Thermodynamics*, North Holland, Amsterdam, 1962.

[19] L. Onsager, Reciprocal Realtions in Irreversible Processes, Phys. Rev. **37**, 405 (1931).

[20] P. Glansdorff and I. Prigogine, *Thermodynamic Theory of Structure, Stability, and Fluctuations*, Wiley-Interscience, New York, 1971.

[21] L. Tisza, in *Thermodynamics: History and Philosophy*, edited by K. Martinas, L. Ropolyi, and P. Szegedi, pages 515–522, World Scientific, Singapore, 1991.

[22] C. Truesdell, *Rational Thermodynamics*, McGraw-Hill, New York, 1985, [2nd enlarged edition (Springer, Berlin, 1988)].

[23] P. T. Landsberg, Foundations of Thermodynamics, Rev. Mod. Phys. **28**, 363 (1956).

[24] B. Chan-Eu, *Kinetic Theory of Irreversible Thermodynamics*, Wiley, New York, 1992.

[25] D. Jou, J. Casas-Vázquez, and G. Lebon, *Extended Irreversible Thermodynamics*, Springer, Berlin, 1993 and 1996, first and second enlarged editions respectively.

[26] R. V. Velasco and L. S. García-Colín, The Kinetic Foundations of Non-Local Nonequilibrium Thermodynamics, J. Non-Equilib. Thermodyn. **18**, 157 (1993).

[27] I. Gyarmati, On the Wave Approach to Thermodynamics and some Problems of Non-Linear Theories, J. Nonequil. Thermodyn. **2**, 233 (1977).

[28] M. Grmela, Dynamics and Thermodynamics of Complex Fluids: I. Development of a General Formalism, J. Chem. Phys. **56**, 6620 (1997).

[29] N. Bernardes, Thermodynamics and Complementarity, Physica A **260**, 186 (1998).

[30] A. Hobson, Irreversibility and Information in Mechanical Systems, J. Chem. Phys. **45**, 1352 (1966).

[31] I. Prigogine, Le Domaine de Validité de la Thermodynamique des Phenomenes Irreversibles, Physica **15**, 272 (1949).

[32] J. Meixner, Thermodynamic of Irreversible Processes Has Many Faces, in *Irreversible Processes of Continuum Mechanics*, edited by H. Parkus and L. Sedov, Springer, Wien, 1968.

[33] I. Müller, Zum Paradoxon der Wärmerleitungstheorie, Zeit. Phys. **198**, 329 (1967).

[34] R. S. Rivlin, Forty Years of Nonlinear Continuum Mechanics, in *Proc. IX Int. Congress on Rheology*, edited by C. R.-L. A. Garcia-Rejón, Universidad Autónoma Metropolitana, México, 1984.

[35] J. C. Maxwell, On the Dynamical Theory of Gases, Phil. Trans. Roy. Soc. (London) **157**, 49 (1867).

[36] C. Cattaneo, Sur Une Forme de L'equation de la Chaleur Eliminant Le Paradoxe D'une Propagation Instantaneè, Comp. Rend. Acad. Sci. (Paris) **247**, 431 (1958).

[37] P. Vernotte, Les Paradoxes de la Théorie de L'equation de la Chaleur, Comp. Rend. Acad. Sci. (Paris) **246**, 3154 (1958).

[38] R. E. Nettleton, Relaxation Theory of Heat Conduction in Liquids, Phys. Fluids **3**, 216 (1960).

[39] R. E. Nettleton, Early Applications of Extended Irreversible Thermodynamics, in *Recent Developments in Nonequilibrium Thermodynamics*, edited by J. Casas-Vázquez, D. Jou, and G. Lebon, Springer, Berlin, 1983.

[40] R. E. Nettleton, Nonlinear Reciprocity and the Maximum Entropy Formalism, Physica A **158**, 672 (1989).

[41] R. Nettleton, Generalized Grad-Type Foundations for Nonlinear Extended Thermodynamics, Phys. Rev. A **42**, 4622 (1990).

[42] R. E. Nettleton, Statistical Comparison of Local and Nonlocal Extended Thermodynamics, S. Afr. J. Phys. **14**, 27 (1991).

[43] I. Müller, *Thermodynamics*, Pitman, London, 1985.

[44] I. Müller, Extended Thermodynamics: Past. Present, Future, in *Recent Devlopments in Nonequilibrium Thermodynamics*, edited by J. Casas-Vázquez, D. Jou, and G. Lebon, Springer, Berlin, 1983.

[45] I. Müller and G. Ruggieri, *Extended Thermodynamics*, Springer, Berlin, 1993.

[46] J. Lambermont and G. Lebon, On a Generalization of Gibbs Equation for Heat Conduction, Phys. Rev. Lett. A **42**, 499 (1973).

[47] D. Jou, J. Casas-Vázquez, and G. Lebon, A Dynamical Interpretation of Second Order Constitutive Equations of Hydrodynamics, J. Nonequil. Thermodyn. **4**, 349 (1979).

[48] D. Jou and J. Casas-Vázquez, Fluctuations of Dissipative Fluxes and the Onsager-Machlup Function, J. Nonequil. Thermodyn. **5**, 91 (1980).

[49] J. Casas-Vázquez and D. Jou, An Approach to Extended Irreversible Thermodynamics: Fluctuation Theory, in *Recent Developments in Nonequilibrium Thermodynamics*, edited by J. Casas-Vázquez, D. Jou, and G. Lebon, Springer, Berlin, 1983.

[50] G. Lebon, An Approach to Extended Irreversible Thermodynamics, in *Recent Developments in Nonequilibrium Thermodynamics*, edited by J. Casas-Vázquez, D. Jou, and G. Lebon, Springer, Berlin, 1983.

[51] G. Lebon, From Classical Irreversible Thermodynamics to Extended Irreversible Thermodynamics, Acta Physica Hungarica **66**, 241 (1989), (Festschrift in honor of I. Gyarmati).

[52] L. S. García-Colín and V. Micenmacher, Some thoughts about the nonequilibrium temperature, Molecular Physics **88**, 399 (1996).

[53] L. S. García-Colín, R. Rodriguez, M. L. de Haro, D. Jou, and J. Casas-Vázquez, On the Foundations of Extended Irreversible Thermodynamics, J. Stat. Phys. **37**, 465 (1984).

[54] L. S. García-Colín, Extended Irreversible Thermodynamics and Chemical Kinetics, in *Recent Developments in Nonequilibrium Thermodynamics*, edited by J. Casas-Vázquez, D. Jou, and G. Lebon, Springer, Berlin, 1983.

[55] L. S. García-Colín, Some Views on Extended Irreversible Thermodynamics, Acta Physica Hungarica **66**, 79 (1989), (Festschrift in honor of I. Gyarmati).

[56] D. Jou, J. Casas-Vázquez, and G. Lebon, Extended Irreversible Thermodynamics, Rep. Prog. Phys. **51**, 1105 (1988).

[57] L. S. García-Colín, Extended Non-Equilibrium Thermodynamics, Scope and Limitations, Revista Mexicana de Fisica **34**, 344 (1988).

[58] L. S. García-Colín and F. Uribe, Extended Irreversible Thermodynamics Beyond the Linear Regime: A Critical Overview, J. Non-Equilib. Thermodyn. **16**, 89 (1991).

[59] G. Lebon, D. Jou, and J. Casas-Vázquez, Questions and Answers About a Thermodynamic Theory of the Third Type, Contemp. Phys. **33**, 41 (1992).

[60] D. Jou, J. Casas-Vázquez, and G. Lebon, Extended Irreversible Thermodynamics: An Overview of Recent Bibliography, J. Nonequil. Thermodyn. **17**, 383 (1992).

[61] J. Casas-Vázquez, D. Jou, and G. Lebon, editors, *Recent Developments in Nonequilibrium Thermodynamics,* Springer, Berlin, 1983.

[62] L. Boltzmann, *Lecture in Gas Theory,* Univ. California Press, Berkeley, California, 1964.

[63] J. M. Ziman, *Electrons and Phonons,* Clarendon, Oxford, 1960.

[64] J. G. Ramos, A. R. Vasconcellos, and R. Luzzi, A Classical Approach in Predictive Statistical Mechanics: A Generalized Boltzmann Formalism, Fortschr. Phys./Prog. Phys. **43**, 265 (1995).

[65] J. G. Ramos, A. R. Vasconcellos, and R. Luzzi, A Generalized Nonlinear Quantum Boltzmann-Like Transport Theory in a Nonequilibrium Ensemble Formalism, Physica A, in press.

[66] G. W. Ford and G. E. Uhlenbeck, in *Lectures in Statistical Mechanics,* edited by M. Kac, Am. Math. Soc., Providence, Rhode Island, 1963, p. 120 et seq.

[67] E. T. Jaynes, Gibbs vs. Boltzmann Entropy, Am. J. Phys. **33**, 391 (1965).

[68] H. Grad, Principles of the Kinetic Theory of Gases, in *Handbuch der Physik XII,* edited by S. Flügge, pages 205–294, Springer, Berlin, 1958.

[69] L. García-Colín and G. Fuentes-Martinez, A Kinetic Derivation of Extended Irreversible Thermodynamics, J. Stat. Phys. **29**, 387 (1982).

[70] H. Mori, I. Oppenheim, and J. Ross, Some Topics in Quantum Statistics, in *Studies in Statistical Mechanics I,* edited by J. de Boer and G. E. Uhlenbeck, pages 217–298, North Holland, Amsterdam, 1962.

[71] E. T. Jaynes, Reprinted Articles and Additional Notes, in *E. T. Jaynes Papers on Probability, Statistics, and Statistical Physics,* edited by R. D. Rosenkrantz, Dordrecht, Reidel, 1983.

[72] B. Robertson, Application of Maximum Entropy to Nonequilibrium Statistical Mechanics, in *The Maximum Entropy Formalism*, edited by M. Tribus and R. D. Levine, pages 289–320, MIT Press, Cambridge, MA, 1978.

[73] T. W. Grandy, Principle of Maximum Entropy and Irreversible Processes, Phys. Rep. **62**, 175 (1980).

[74] R. Kubo, in *Lectures in Theoretial Physics*, edited by W. Brittin, Wiley, New York, 1959.

[75] R. Kubo, The Fluctuation-Dissipation Theorem, in *Many-Body Problems*, edited by S. F. Edwards, Benjamin, New York, 1969.

[76] H. Mori, Relaxation Phenomena Near the Critical Points, in *Many-Body Theory*, edited by R. Kubo, Syokabo, Tokyo, 1966.

[77] H. Mori, Transport, Collective Motion, and Brownian Motion, Progr. Theor. Phys. (Japan) **33**, 423 (1965).

[78] H. Mori, A Continued-Fraction Representation of the Time-Correlation Functions, Prog. Theor. Phys. (Japan) **34**, 399 (1965).

[79] R. Zwanzig, Time-Correlation Functions and Transport Coefficients in Statistical Mechanics, in *Annual Review of Physical Chemistry*, volume 16, pages 67–102, Academic Press, New York, 1965.

[80] R. Zwanzig, Statistical Mechanics of Irreversibility, in *Lectures in Theoretical Physics*, Vol. 3, edited by W. E. Brittin, B. W. Downs, and J. Downs, Wiley-Interscience, New York, 1961.

[81] R. Zwanzig, Memory Effects in Irreversible Thermodynamics, Phys. Rev. **124**, 983 (1961).

[82] L. S. García-Colín and J. L. D. Rio, A Unified Approach for Deriving Kinetic Equations in Nonequilibrium Statistical Mechanics, J. Stat. Phys. **16**, 235 (1977).

[83] H. Grabert, *Projection Operators Techniques in Nonequilibrium Statistics*, Springer, Berlin, 1981.

[84] I. Prigogine, Structure, Dissipation, and Life, in *From Theoretical Physics to Biology*, edited by M. Marois, North Holland, Amsterdam, 1969.

[85] I. Prigogine, G. Nicolis, and A. Babloyantz, Thermodynamics of Evolution I, Phys. Today **25**(11), 23 (1972).

[86] I. Prigogine, G. Nicolis, and A. Babloyantz, Thermodynamics of Evolution II, Phys. Today **25**(12), 38 (1972).

[87] G. Nicolis and I. Prigogine, *Exploring Complexity*, Freeman, New York, 1989.

[88] O. Penrose, Foundations of Statistical Mechanics, Rep. Prog. Phys. **42**, 1938 (1979).

[89] R. Kubo, Oppening Address at the Oji Seminar, Prog. Theor. Phys. (Japan) **Suppl. 64**, 1 (1978).

[90] R. Zwanzig, Where do we go from here?, in *Perspectives in Statistical Physics*, edited by H. J. Raveché, pages 123–124, North Holland, Amsterdam, 1981.

[91] L. S. García-Colín, Selected Topics in Nonequilibrium Phenomena, in *Monografias Em Física*, Instituto de Física "Gleb Wataghin" — UNICAMP, Campinas, SP, 1984.

[92] L. S. García-Colín and M. S. Green, The Chapman-Enskog Solution of the Generalized Boltzmann Equation, Physica **32**, 450 (1966).

[93] N. N. Bogoliubov, *Lectures in Quantum Statistics*, volume 1 and 2, Gordon and Breach, New York, 1967 and 1970 respectively.

[94] H. J. Kreuzer, *Nonequilibrium Thermodynamics and its Statistical Foundations*, Clarendon, Oxford, 1981.

[95] H. Mori, Statistical-Mechanical Theory of Transport in Fluids, Phys. Rev. **112**, 1829 (1958).

[96] D. Forster, *Hydrodynamic Fluctuations, Broken Symmetry, and Correlation Functions*, Benjamin, Readings, MA, 1975.

[97] L. L. Lebowitz and E. W. Montroll, Preface, in *Studies in Statistical Mechanics X. Nonequilibrium Phenomena I: The Boltzmann Equation*, North Holland, Amsterdam, 1983.

[98] J. A. McLennan, Nonlinear Effects in Transport Theory, Phys. Fluids **4**, 1319 (1961).

[99] S. V. Peletminskii and A. A. Yatsenko, Contribution to the Quantum Theory of Kinetic and Relaxation Processes, Soviet Phys. JETP **26**, 773 (1968), [Zh. Ekps. Teor. Fiz. **53**, 1327 (1967)].

[100] D. N. Zubarev, *Nonequilibrium Statistical Thermodynamics*, Consultants Bureau, New York, 1974, [Neravnovesnaia Statisticheskaia Termodinamika (Izd. Nauka, Moscow, 1971)].

[101] R. Luzzi and A. R. Vasconcellos, On the Nonequilibrium Statistical Operator Method, Fortschr. Phys./Prog. Phys. **38**, 887 (1990).

[102] E. T. Jaynes, Macroscopic Prediction, in *Complex Systems: Operational Approaches*, edited by H. Haken, Springer, Berlin, 1985.

[103] E. T. Jaynes, Predictive Statistical Mechanics, in *Frontiers of Nonequilibrium Statistical Physics*, edited by G. T. Moore and M. O. Scully, pages 33–55, Plenum, New York, 1986.

[104] E. T. Jaynes, Clearing up Mysteries — The Original Goal, in *Maximum Entropy and Bayesian Methods*, edited by J. Skilling, pages 1–27, Kluwer, Dordrecht, 1989.

[105] E. T. Jaynes, Notes on present status and future prospects, in *Maximum Entropy and Bayesian Methods*, edited by W. T. Grandy and L. H. Schick, pages 1–13, Kluwer, Dordrecht, 1990.

[106] E. T. Jaynes, Information Theory and Statistical Mechanics I, Phys. Rev. **106**, 620 (1957).

[107] C. E. Shannon and W. Weaver, *The Mathematical Theory of Communication.*, Univ. Illinois Press, Urbana, 1948.

[108] R. Luzzi, A. R. Vasconcellos, and J. G. Ramos, On the Statistical Foundations of Irreversible Thermodynamics, Fortschr. Phys./Prog. Phys. **47**, 401 (1999).

[109] R. Luzzi and A. R. Vasconcellos, The Basic Principles of Irreversible Thermodynamics in the Context of an Informational-Statistical Approach, Physica A **241**, 677 (1997).

[110] J. G. Ramos, A. R. Vasconcellos, and L. S. García-Colín, A Thermo-Hydrodynamic Theory Based on Informational Statistical Thermodynamics, Braz. J. Phys. **27**, 585 (1997).

[111] W. T. Grandy, *Principles of Statistical Mechanics: Equilibrium Theory*, volume 1, Reidel, Dordrecht, 1987.

[112] W. T. Grandy, *Principles of Statistical Mechanics: Nonequilibrium Phenomena*, volume 2, Reidel, Dordrecht, 1988.

[113] A. I. Akhiezer and S. V. Peletminskii, *Methods of Statistical Physics*, Pergamon, Oxford, 1981.

[114] J. A. McLennan, Statistical Theory of Transport Processes, in *Advances in Chemical Physics*, volume 5, pages 261–317, Academic, New York, 1963.

[115] J. G. Kirkwood, The Statistical Mechanical Theory of Transport Processes, J. Chem. Phys. **14**, 180 (1946).

[116] M. S. Green, Markoff Random Processes and the Statistical Mechanics of Time-Dependent Phenomena I, J. Chem. Phys. **20**, 1281 (1952).

[117] D. N. Zubarev and V. P. Kalashnikov, Extremal Properties of the Nonequilibrium Statistical Operator, Theor. Math. Phys. **1**, 108 (1970).

[118] B. Robertson, Equations of Motion in Nonequilibrium Statistical Mechanics, Phys. Rev. **144**, 151 (1966).

[119] V. P. Kalashnikov, Equations of Motion, Green's Functions, and Thermodynamic Relations in Theories of Linear Relaxation, Theor. Math. Phys. **35**, 362 (1978).

[120] V. P. Vstovskii, Projection Formalism in the Theory of Irreversible Processes, Theor. Math. Phys. **21**, 1214 (1975).

[121] M. V. Sergeev, Generalized Transport Equations in the Theory of Irreversible Processes, Theor. Math. Phys. **21**, 1234 (1975).

[122] L. Lauck, A. R. Vasconcellos, and R. Luzzi, A Nonlinear Quantum Transport Theory, Physica A **168**, 789 (1990).

[123] S. Hawkings, The Arrow of Time, in *1990 - Yearbook of Science and Future*, Encyclopaedia Britannica, Chicago, 1989.

[124] R. Lestiene, Entropy, Mechanical Time, and Cosmological Arrow, Scientia **113**, 313 (1980).

[125] I. Prigogine, *From Being to Becoming*, Freeman, San Francisco, 1980.

[126] I. Prigogine and I. Stengers, *Entre le Temps et l'Eternité*, Fayard, Paris, 1988.

[127] I. Prigogine and I. Stengers, *Order out of the Chaos: Man's New Dialogue with Nature*, Bantam, New York, 1984.

[128] P. V. Coveney and R. Highfield, *The Arrow of Time*, Fawcett Columbine, New York, 1990.

[129] N. N. Bogoliubov, Some Topics in Quantum Statistics, in *Studies in Statistical Mechanics I*, edited by J. de Boer and G. E. Uhlenbeck, North Holland, Amsterdam, 1962.

[130] G. E. Uhlenbeck, The Boltzmann Equation, in *Lectures in Statistical Mechanics*, edited by M. Kac, pages 183–203, Am. Math. Soc, Providence, RI, 1963.

[131] L. L. Buishvili and M. D. Zviadadze, On the Quasi-Thermodynamic Theory of Magnetic Relaxation, Physica **59**, 697 (1972).

[132] A. R. Vasconcellos, A. C. Algarte, and R. Luzzi, On the Question of Bogoliubov Hierarchy of Relaxation Times: An Example from Semiconductor Physics, Physica A **166**, 517 (1990).

[133] A. R. Vasconcellos, R. Luzzi, and L. S. García-Colín, A Microscopic Approach to IrreversibleThermodynamics I: General Theory, Phys. Rev. A **43**, 6622 (1991).

[134] A. R. Vasconcellos, R. Luzzi, and L. S. García-Colín, A Microscopic Approach to Irreversible Thermodynamics II: An Example from Semiconductor Physics, Phys. Rev. A **43**, 6633 (1991).

[135] J. G. Ramos, A. R. Vasconcellos, and R. Luzzi, A Nonequilibrium Ensemble Formalism: Criterion for Truncation of Description, J. Chem. Phys. (1999), in press.

[136] I. Prigogine, The Statistical Interpretation of Non-Equilibrium Entropy, Acta Physica Austríaca **Suppl. X**, 401 (1973).

[137] S. A. Hassan, A. R. Vasconcellos, and R. Luzzi, The informational-statistical-entropy operator in a nonequilibrium ensemble formalism, Physica A **262**, 359 (1999).

[138] N. N. Bogoliubov, *Lectures in Quantum Statistics II*, Gordon and Breach, New York, 1970.

[139] E. S. Freidkin and R. E. Nettleton, Consistency Between Maximum-Entropy Formalism and H-Theorem, Nuovo Cimento B **104**, 597 (1989).

[140] J. E. Mayer, Approach to Thermodynamic Equilibrium, J. Chem. Phys **34**, 1207 (1961).

[141] J. L. Lebowitz, H. L. Frisch, and E. Helfand, Nonequilibrium Correlation Functions in a Fluid, Phys. Fluids **3**, 325 (1960).

[142] R. M. Lewis, A Unifying Principle in Statistical Mechanics, J. Math. Phys. **8**, 1448 (1967).

[143] J. Karkheck, H. V. Beijeren, M. S. I, and G. Stell, Kinetic Theory and H-Theorem for a Dense Square-Well Fluid, Phys. Rev. A **32**, 2517 (1985).

[144] R. C. Castello, E. Martina, M. L. de Haro Amd J. Karkheck, and G. Stell, Linearized Kinetic-Variational Theory and Short-Time Kinetic Theory, Phys. Rev. A **39**, 3106 (1989).

[145] D. N. Zubarev, Nonequlibrium Statistical Mechanics, in reference [100], see Chapter IV, Section 22.

[146] R. Nettleton, Information Theoretic Extended Entropy for Steady Heat Conduction in Dense Fluids, J. Phys. A **21**, 3939 (1988).

[147] R. E. Nettleton, Corrections to Maximum Entropy for Steady-State Conduction, J. Phys. A **22**, 5281 (1989).

[148] R. E. Nettleton and E. S. Freidkin, Nonlinear Reciprocity and the Maximum Entropy Formalism, Physica A **158**, 672 (1989).

[149] R. Nettleton, Evaluation of Correlations Defining Coefficients in Nonequilibrium Thermodynamics, J. Phys. Soc. Japan **61**, 3103 (1992).

[150] N. G. V. Kampen, Chapman-Enskog as an Application of the Method of Eliminating Fast Variables, J. Stat. Phys. **46**, 709 (1987).

[151] A. C. Algarte, A. R. Vasconcellos, and R. Luzzi, Kinetic of Hot Elementary Excitations in Photoexcited Polar Semiconductors, Phys. Stat. Sol. (b) **173**, 487 (1992).

[152] A. R. Vasconcellos, A. C. Algarte, and R. Luzzi, Diffusion of Photoinjected Carriers in Plasma in Nonequilibrium Semiconductors, Phys. Rev. B **48**, 10873 (1993).

[153] R. Luzzi and A. R. Vasconcellos, Ultrafast Transient Response of Nonequilibrium Plasma in Semiconductors., in *Semiconductor Processes Probed by Ultrafast Laser Spectroscopy*, edited by R. R. Alfano, volume 1, pages 135–169, Academic, New York, 1984.

[154] R. Luzzi, A. R. Vasconcellos, and A. S. Esperidião, Propagation of Damped Hydrodynamic Modes in a Photoinjected Plasma in Semiconductors, Phys. Rev. B **52**, 5021 (1995).

[155] A. R. Vasconcellos, R. Luzzi, D. Jou, and J. Casas-Vázquez, Second Sound Wave in Photoinjected Plasma in Semiconductors, Phys. Rev. B **52**, 5030 (1995).

[156] J. Madureira, A. Vasconcellos, and R. Luzzi, A Nonequilibrium Statistical Grand-Canonical Ensemble: Description in Terms of Flux Operators, J. Chem. Phys. **109**, 2099 (1998).

[157] S. V. Peletminskii and A. I. Sokolovskii, Flux Operators of Physical Variables and the Method of Quasi-Averages, Theor. Math. Phys. **18**, 85 (1974).

[158] A. R. Vasconcellos, R. Luzzi, D. Jou, and J. Casas-Vázquez, Thermal Waves in an Extended Hydrodynamic Approach, Physica A **212**, 369 (1995).

[159] A. R. Vasconcellos, R. Luzzi, and L. S. García-Colín, A Microscopic Approach to Irreversible Thermodynamics III: Generalized Constitutive Equations, J. Non-Equilib. Thermodyn. **20**, 103 (1995).

[160] A. R. Vasconcellos, R. Luzzi, and L. S. García-Colín, A Microscopic Approach to Irreversible Thermodynamics IV: Diffusive and Wave Thermal Motion, J. Non-Equilib. Thermodyn. **20**, 119 (1995).

[161] A. R. Vasconcellos, R. Luzzi, and L. S. García-Colín, A Microscopic Approach to Irreversible Thermodynamics V: Memory-Dependent Constitutive Equations, J. Mod. Phys. **9**, 1933 (1995).

[162] A. R. Vasconcellos, R. Luzzi, and L. S. García-Colín, A Microscopic Approach to Irreversible Thermodynamics VI: Equations of Evolution nonlinear in the Fluxes in Informational Statistical Thermodynamics, J. Mod. Phys. **9**, 1945 (1995).

[163] A. R. Vasconcellos, R. Luzzi, and L. S. García-Colín, A Microscopic Approach to Irreversible Thermodynamics VII: Response Function Theory for Thermal Perturbations in Informational Statistical Thermodynamics, Physica A **221**, 478 (1995).

[164] A. R. Vasconcellos, R. Luzzi, and L. S. García-Colín, A Microscopic Approach to Irreversible Thermodynamics VIII: Diffusion and Mobility and Generalized Einstein Relation, Physica A **221**, 495 (1995).

[165] J. G. Ramos, A. R. Vasconcellos, and R. Luzzi, Microscopic Approach to Irreversible Thermodynamics IX: A Generalized Hydrodynamic Theory, as yet unpublished.

[166] M. A. Tenan, A. R. Vasconcellos, and R. Luzzi, Statistical Foundations of Generalized Nonequilibrium Thermodynamics, Forstchr. Phys./Prog. Phys. **47**, 1 (1997).

[167] J. L. D. Rio and L. S. García-Colín, Increase-in-Entropy Law, Phys. Rev. E **48**, 819 (1993).

[168] L. Brillouin, *Science and Information Theory*, Academic Press, New York, 1962.

[169] G. Nicolis and I. Prigogine, *Self-organization in Nonequilibrium Systems*, Wiley-Interscience, New York, 1977.

[170] A. R. Vasconcellos and R. Luzzi, On the Statistical Thermodynamics of a Model of Nonequilibrium Semiconductors, Physica A **180**, 182 (1992).

[171] O. Gurel and O. E. Rössler, editors, *Bifurcation Theory and Applications in Scientific Disciplines*, Vol 316 of the Annals of the New York Academy of Sciences, New York, 1979.

[172] G. Nicolis, Dissipative Systems, Rep. Prog. Phys. **49**, 873 (1986).

[173] D. Jou and J. Casas-Vázquez, Possible Experiment to Check the Reality of a Nonequilibrium Temperature, Phys. Rev. A **45**, 8371 (1992).

[174] R. Luzzi, A. R. Vasconcellos, J. Casas-Vázquez, and D. Jou, On the Selection of the State Space in Nonequlibrium Thermodynamics, Physica A **248**, 111 (1998).

[175] R. Luzzi and A. R. Vasconcellos, Response Function Theory for Far-from-Equilibrium Systems, J. Stat. Phys. **23**, 539 (1980).

[176] A. R. Vasconcellos, R. Luzzi, D. Jou, and J. Casas-Vázquez, Thermodynamic Variables in the Context of a Nonequilibrium Ensemble Formalism, J. Chem. Phys. **107**, 7383 (1997).

[177] B. Doubrovine, S. Novikov, and A. Fomenko, *Géometrie Contemporaine*, volume 1, MIR, Moscow, 1985.

[178] R. Courant and D. Hilbert, *Methods of Mathematical Physics*, Wiley-Interscience, New York, 1953.

[179] S. P. Heims and E. T. Jaynes, Theory of Gyromagnetic Effects and some Related Magnetic Phenomena, Rev. Mod. Phys. **34**, 143 (1962), Subsection b, pp. 148–150, and Appendix B, p. 164 (It should be

noticed a misprint in the third line of their Eq. (B1) which must end in x^{n-1}).

[180] R. Luzzi, A. R. Vasconcellos, and J. G. Ramos, Foundations of a Nonequilibrium Ensemble Formalism, in *Fundamental Theories of Physics Series*, edited by A. V. der Merwe, Kluwer Academic, Dordrecht, In preparation.

[181] R. Balian, Y. Alhassid, and H. Reinhardt, Dissipation in Many-Body Systems: A Geometric Approach Based in Information Theory, Phys. Rep. **131**, 1 (1986).

[182] N. W. Aschroft and N. D. Mermin, *Solid State Physics*, Holt, Reinhart, and Winston, New York, 1976.

[183] E. Conwell, High Field Transport in Semiconductors, in *Solid State Physics*, edited by F. Seitz, D. Turnbull, and H. Ehrenreich, volume 1, pages 1–293, Academic Press, New York, 1967, Suppl. 9.

[184] J. Madureira, A. Vasconcellos, R. Luzzi, and L. Lauck, Markovian Kinetic Equations in a Nonequilibrium Statistical Ensemble Formalism, Phys. Rev. E **57**, 3637 (1998).

[185] P. M. Morse and H. Feschbach, *Methods of Theoretical Physics*, McGraw-Hill, New York, 1953.

[186] R. Luzzi, M. A. Scarparo, J. G. Ramos, A. R. Vasconcellos, M. L. Barros, Z. Zhyiao, and A. Kiel, Informational Statistical Thermodynamics and Thermal Laser Stereolithography, J. Non-Equilib. Thermodyn. **22**, 197 (1997).

[187] J. R. Madureira, A. R. Vasconcellos, R. Luzzi, J. Casas-Vázquez, and D. Jou, Evolution of Dissipative Processes in a Statistical Thermodynamic Approach I: Generalized Mori-Heisenberg-Langevin Equations, J. Chem. Phys. **108**, 7568 (1998).

[188] J. R. Madureira, A. R. Vasconcellos, R. Luzzi, J. Casas-Vázquez, and D. Jou, Evolution of Dissipative Processes in a Statistical Thermodynamic Approach II: Thermodynamic Properties of a Fluid of Bosons, J. Chem. Phys. **108**, 7580 (1998).

[189] R. Luzzi, A. R. Vasconcellos, J. Casas-Vázquez, and D. Jou, Characterization and Measurement of a Nonequilibrium Temperature-Like Variable in Irreversible Thermodynamics, Physica A **234**, 699 (1997).

[190] A. R. Vasconcellos, R. Luzzi, D. Jou, and J. Casas-Vázquez, A Nonequilibrium Temperature-Like Variable in Informational-Statistical Thermodynamics, J. Chem. Phys. (1998), submitted.

[191] A. C. Algarte, A. R. Vasconcellos, and R. Luzzi, Ultrafast Phenomena in the Photoinjected Plasma in Semiconductors, Braz. J. Phys. **26**, 543 (1996).

[192] P. Motizuke, C. A. Arguello, and R. Luzzi, Effects of Excited Electron States Lifetime on Gain Spectra of EHL in CdS, Solid State Commun. **23**, 617 (1977).

[193] N. Oreskes, H. Shrader-Frechette, and K. Beltz, Verification, Validation, and Confirmation of Numerical Models in the Earth Sciences, Science **263**, 641 (1994).

[194] S. J. Gould, *Dinosaur in a Haystack*, Random House, New York, 1995.

[195] R. Jancel, *Foundations of Classical and Quantum Satistical Mechanics*, Pergamon, Oxford, 1963.

[196] M. Criado-Sancho and J. E. Llebot, Behavior of Entropy in Hyperbolic Heat Conduction, Phys. Rev. E **47**, 4104 (1993).

[197] J. Ramos, A. Vasconcellos, and R. Luzzi, On the Relationship of Different Approaches in Statistical Thermodynamics of Dissipative Systems, IFGW-Unicamp Internal-Reportl (1998), unpublished.

[198] G. Nicolis, Physics of Far-From-Equilibrium Systems and Self-Organization, in *The New Physics*, edited by P. Davies, pages 316–347, Cambridge Univ. Press, Cambridge, 1989.

[199] G. Careri, *Order and Disorder in Matter*, Benjamin/Cummings, New York, 1984.

[200] G. Nicolis, Introductory remarks: Thermodynamics today, Physica A **213**, 1 (1995).

[201] A. R. Vasconcellos, A. C. Algarte, and R. Luzzi, Ultrafast Phenomena in the Photoinjected Plasma in Semiconductors, Braz. J. Phys **26**, 543 (1996).

[202] E. T. Jaynes, The Evolution of Carnot's Principle, in *Maximum Entropy And Bayesian Methods in Science and Engineering*, edited by G. J. Erickson and C. R. Smith, volume 1, pages 267–281, Kluwer, Amsterdam, 1988.

[203] L. Rosenfeld, A Question of Physics, in *Conversations in Physics and Biology*, edited by P. Buckley and D. Peat, Univ. Toronto Press, Toronto, 1979.

[204] L. Rosenfeld, Questions of Irreversibility and Ergodicity, in *Proceedins of the Int. School Phys. "Enrico Fermi", Course XIV*, edited by P. Caldirola, pages 1–20, New York, 1960, Academic.

[205] L. Rosenfeld, On the Foundations of Statistical Thermodynamics, Acta Phys. Polonica **14**, 3 (1955).

[206] R. Luzzi, J. G. Ramos, and A. R. Vasconcellos, Rosenfeld-Prigogine's Complementarity of Description in the Context of Informational Statistical Thermodynamics, Phys. Rev. E **57**, 244 (1998).

[207] C. Kittel, Temperature Fluctuation: An Oxymoron, Physics Today **41**(5), 93 only (1988).

[208] R. K. Pathria, *Statistical Mechanics*, Addison-Wesley, Readings, MA, 1972.

[209] N. Bohr, On the Notions of Causality and Complementarity, Dialectica **2**, 312 (1948).

[210] L. D. Landau and E. M. Lifshitz, *Statistical Physics*, Addison-Weslwy, Reading, Massashusetts, 1958, second enlarged edition,1969.

[211] D. N. Zubarev, V. N. Morozov, and G. Röpke, *Statistical Mechanics of Nonequilibrium Processes: Basic Concepts, Kinetic Theory*, volume 1, Akademie Verlag Wiley VCH, Berlin, 1996.

[212] V. P. Kalashnikov, Linear Relaxation Equations in the Nonequilibrium Statistical Operator Method, Theor. Math. Phys. **34**, 263 (1978), Teor. Mat. Fiz. **34**, 412 (1978).

[213] S. Kullback, *Information Theory and Statistics*, Wiley, New York, 1951.

[214] F. Schlögl, Stochastic Measures in Nonequilibrium Thermodynamics, Phys. Rep. **62**, 267 (1980).

[215] J. N. Kapur and H. K. Kesavan, *Entropy Optimization Principles with Applications*, Academic, Boston, 1992.

[216] R. Luzzi, A. R. Vasconcellos, and J. G. Ramos, Statistical Irreversible Thermodynamics in a Nonequilibrium Statistical Ensemble Formalism, Phys. Reports. (1999), submitted.

[217] J. L. D. Rio and L. S. García-Colín, Concept of Entropy for Nonequilibrium States of Closed Many-Body Systems, Phys. Rev. A **43**, 6657 (1991).

[218] N. S. Krylov, *Works on the Foundations of Statistical Mechanics*, Princeton Univ. Press, Princeton, 1979, with an Introduction by D. ter Haar.

[219] H. Jeffreys, *Probability Theory*, Clarendon, Oxford, 1961.

[220] P. W. Anderson, The Reverend Thomas Bayes, Needles In a Haystack, and the Fifth Force, Phys. Today **45**, 9 (jan 1992).

[221] A. J. Garret, Essay Review, Entropy, Contemp. Phys. **33**, 271 (1992).

[222] J. Luczka, Generalized Kinetic Equations with Memory, Phys. Lett A **69**, 393 (1979).

[223] H. Jeffreys, *Scientific Inference*, Cambridge Univ. Press, Cambridge, third edition, 1973.

[224] J. P. Dougherty, Approaches to Non-Equilibrium Statistical Mechanics, in *Maximum Entropy and Bayesian Method*, edited by J. Skilling, pages 131–136, Kluwer, Dordrecht, 1989.

[225] J. P. Dougherty, Explaining Statistical Mechanics, Stud. Phil. Sci. **24**, 843 (1993).

[226] J. P. Dougherty, Foundations of Non-Equilibrium Statistical Mechanics, Phil. Trans. R. Soc. Lond. A **346**, 259 (1994).

[227] J. Bricmont, Science of Chaos or Chaos in Science, in *The Flight from Science and Reason*, The New York Academy of Science, New York, 1996, Annals of the New York Academy of Science, Vol 775.

[228] J. Meixner, The Entropy Problem in Thermodynamic Processes, Rheol. Acta **12**, 465 (1973).

[229] J. Meixner, Entropy and Entropy Production, in *Foundations of Continuum Thermodynamics*, edited by J. J. Delgado, M. N. R. Nina, and J. H. Whitelaw, pages 129–141, McMillan, London, 1974.

Name Index

Index

Printed in the United States
By Bookmasters